SpringerBriefs in Physics

SpringerBriefs in Physics are a series of slim high-quality publications encompassing the entire spectrum of physics. Manuscripts for SpringerBriefs in Physics will be evaluated by Springer and by members of the Editorial Board. Proposals and other communication should be sent to your Publishing Editors at Springer.

Featuring compact volumes of 50 to 125 pages (approximately 20,000–45,000 words), Briefs are shorter than a conventional book but longer than a journal article. Thus, Briefs serve as timely, concise tools for students, researchers, and professionals.

Typical texts for publication might include:

- A snapshot review of the current state of a hot or emerging field
- A concise introduction to core concepts that students must understand in order to make independent contributions
- An extended research report giving more details and discussion than is possible in a conventional journal article
- A manual describing underlying principles and best practices for an experimental technique
- An essay exploring new ideas within physics, related philosophical issues, or broader topics such as science and society

Briefs allow authors to present their ideas and readers to absorb them with minimal time investment. Briefs will be published as part of Springer's eBook collection, with millions of users worldwide. In addition, they will be available, just like other books, for individual print and electronic purchase. Briefs are characterized by fast, global electronic dissemination, straightforward publishing agreements, easy-to-use manuscript preparation and formatting guidelines, and expedited production schedules. We aim for publication 8–12 weeks after acceptance.

Jaroslav Zamastil

Understanding the Path from Classical to Quantum Mechanics

 Springer

Jaroslav Zamastil
Charles University in Prague
Praha 2, Czech Republic

ISSN 2191-5423   ISSN 2191-5431   (electronic)
SpringerBriefs in Physics
ISBN 978-3-031-37372-5   ISBN 978-3-031-37373-2   (eBook)
https://doi.org/10.1007/978-3-031-37373-2

This Springer imprint is published by the registered company Springer Nature Switzerland AG
The registered company address is: Gewerbestrasse 11, 6330 Cham, Switzerland

Preface

To my mind, great works can only be born within the history of their art and as participants in that history. It is only inside history that we can see what is new and what is repetitive, what is discovery and what is imitation; in other words, only inside history can a work exist as a value capable of being discerned and judged. Nothing seems to me worse for art than to fall outside its own history, for it is a fall into the chaos where esthetic values can no longer be perceived. M. Kundera [1]

While almost all textbooks on quantum mechanics mention the old quantum theory of Bohr and Sommerfeld, which was later found to be incorrect due to both inconsistencies within the theory and disagreement with experimental data, there is hardly any educational material on how Heisenberg arrived at his "magical steps" in 1925. As much as the collection of the most important papers with insightful introduction [2] and the thorough historical study [3] must be praised, they hardly represent anything that could be called pedagogical exposition. While the work [4] clarified the missing steps in Heisenberg's 1925 calculation, it did not address how Heisenberg arrived at the principles upon which his calculation was based.

The purpose of this work is not to uncover new historical information but rather to provide a pedagogical approach to understanding the discovery of quantum mechanics. We do not aim to find anything new in historical aspects; in fact, with respect to historical accuracy, we completely relied on standard sources [2, 3]. The goal is not to present a complete historical account, but rather to follow the narrow path that led to the discovery of quantum mechanics. The focus is on the main ideas, not the step-by-step process of pioneers, resulting in a more cartoon history than a real one.

We believe that providing a pedagogical exposition of how quantum mechanics was discovered would be of interest to both beginners and practitioners in the field. For students, it could serve as a supplement to standard quantum mechanics textbooks. In the author's experience, good practitioners are typically interested in the history of their subject, and with good reason: science is a human endeavor, and even the greatest scientists are only human. While the historical approach may not always be the best pedagogical presentation of the ideas, it can certainly aid in both scientific and human understanding.

This essay is structured as follows: Chap. 1 briefly summarizes the component formalism for vector calculus and the notion of the δ-function. While these are well-known topics that have been covered in many places, we include them here to help students and to establish a clear notation. Chapters 2 and 3 provide necessary background for the transition between classical and quantum theory, summarizing parts of classical mechanics and electrodynamics. In Chap. 2, we provide a comprehensive classical theory of the emission, absorption, dispersion, and scattering of light by matter, using the oscillator model proposed by Lorentz, Drude, Planck, and others to describe matter. In Chap. 3, we present an exposition of Hamilton's formulation of classical mechanics, beginning with the Hamilton varia-tional principle and proceeding to derive Hamilton's canonical equations, canonical transformations, and the action-angle variables. Chapter 4 outlines the steps toward finding a correct solution, starting with a summary of several significant, empirically valid formulas, namely the Planck radiation law, the Rydberg-Ritz combination principle, and the Balmer formula for the frequencies of hydrogen spectral lines. We then explain Einstein's 1916 derivation of the Planck radiation law and the Ladenburg-Kramers-Born-Heisenberg theory of dispersion. The portions of this chapter that focus on the canonical perturbation method rely heavily on Hamilton's canonical formalism and can be rather involved. Nonetheless, they are indispensable for understanding the classical dispersion theory and its quantum reformulation, which ultimately led to Heisenberg's decisive breakthrough. Finally, in Chap. 5, we attempt to explain Heisenberg's 1925 "magical steps" that led to the discovery of quantum mechanics. Although we try hard to demystify some of the magic involved, some magic still remains, as it is clear that Heisenberg's 1925 paper represents the work of a genius.

So, the sequence of events that led to the discovery of quantum mechanics and that we aim to cover here is as follows:

- In 1916, Albert Einstein derived the Planck radiation law and introduced coefficients of absorption and emission to describe the interaction of atoms and radiation.
- In 1921, Rudolf Ladenburg related the oscillator strengths of classical dispersion theory to the Einstein absorption and emission coefficients.
- In 1924, Hendrik Kramers extended the Ladenburg formula and Max Born derived the Kramers formula. Born related the Einstein coefficients to the square of the transition amplitudes, which are quantum replacements of Fourier coefficients of classical coordinates.
- In 1925, Werner Heisenberg established the laws that govern these transition amplitudes, and quantum mechanics was born.

The quantum theory had its beginnings in 1900 with Max Planck's attempt to derive his own empirically found radiation law. To reconcile it with experience, he had to postulate that the energies of his material oscillators have to be quantized. In 1905, Albert Einstein critically analyzed Planck's derivation and proposed that the energy of the radiation, rather than the energy of the matter oscillators, is quantized. He then showed that this hypothesis explains the photoelectric effect. In 1913, Niels Bohr published his atomic model, assuming the validity of classical description of motion, but applying to it his quantum conditions. He subsequently provided, based on his model, a qualitative explanation of the regularities of Mendeleev periodic table of elements. Arnold Sommerfeld then reformulated the Bohr theory in terms of action-angle variables and demonstrated that relativistic corrections explain the fine structure of atoms. Although these early developments were significant, they are not directly relevant to the subject of this essay, and they are already covered in many undergraduate textbooks and popular books. Similarly, the developments leading to Schrödinger wave mechanics and its formulation are well covered in existing literature. Moreover, this development is somewhat easier to understand than the development leading to quantum mechanics. While the original Schrödinger papers are highly regarded, they are not considered "magical."

Prerequisites

We assume knowledge of classical electrodynamics and theoretical mechanics at the undergraduate level. For closer acquaintance with these subjects, see, e.g., [5, 6] for electrodynamics and [7] for theoretical mechanics. Anyone, who has mastered the material of the standard undergraduate courses on classical electrodynamics and theoretical mechanics, should understand this essay without much trouble.

Praha 2, Czech Republic   Jaroslav Zamastil

Acknowledgements

The author expresses gratitude to Tereza Uhlířová for carefully reviewing the manuscript and rectifying the most glaring English errors. Special appreciation is extended to Vojtěch Kapsa for affording the author the chance to deliver lectures on the book's subject matter during the "Seminar on Quantum Theory" in the Academic Year 2022/23. Additionally, thanks are owed to Sam Harrison, the editor at Springer, for providing invaluable assistance with the manuscript.

Contents

Chapter 1
Mathematical Preliminaries

In this chapter, we summarize our units, conventions, component formalism, notion of δ-functions, and some results concerning the transition from Cartesian to spherical coordinates.

1.1 Units

We work in natural units $\hbar = c = 1$. As the value of the reduced Planck constant and speed of light can be, by change of units, rescaled to one, their values in SI only serve to relate the atomic scale to everyday scale. Otherwise, they are of no significance. In natural units, the fine structure constant equals

$$\alpha = \frac{e^2}{4\pi} \simeq \frac{1}{137.036}.$$ (1.1)

1.2 The Summation Convention

- The scalar product is denoted by a centered dot, i.e., $\vec{a} \cdot \vec{b}$.
- The components of three vectors carry a Latin index ($i, j, k, \ldots$).
- The Einstein summation convention is used throughout the book: if two same indices appear, they are summed over; for instance,

$$a_i b_i = a_1 b_1 + a_2 b_2 + a_3 b_3.$$

© The Author(s), under exclusive license to Springer Nature Switzerland AG 2023
J. Zamastil, *Understanding the Path from Classical to Quantum Mechanics*,
SpringerBriefs in Physics, https://doi.org/10.1007/978-3-031-37373-2_1

1.3   The Component Formalism

For vectors and their differentiations, we use customary notation, see, e.g., [5]. The dot over the quantity means, as usual, differentiation with respect to time. The scalar product may be written in components by means of the above-mentioned Einstein summation convention as

$$\vec{A} \cdot \vec{B} = A_j B_j = \delta_{ij} A_i B_j \,, \qquad \nabla \cdot \vec{A} = \frac{\partial A_i}{\partial x_i} \,, \qquad \nabla^2 = \frac{\partial^2}{\partial x_i \partial x_i} \,, \qquad (1.2)$$

where the Kronecker symbol

$$\delta_{ij} = \begin{cases} 1 & (i = j) \,, \\ 0 & (i \neq j) \,. \end{cases}$$

The vector product may be written in components as

$$(\vec{A} \times \vec{B})_i = \epsilon_{ijk} A_j B_k \,, \qquad (\nabla \times \vec{A})_i = \epsilon_{ijk} \frac{\partial A_k}{\partial x_j} \,,$$

where the Levi–Civita symbol

$$\epsilon_{123} = \epsilon_{231} = \epsilon_{312} = 1 \,,$$
$$\epsilon_{213} = \epsilon_{132} = \epsilon_{321} = -1 \,,$$

and

$$\epsilon_{ijk} = 0$$

when any two indices ij, jk, or ik take the same value.

The identities of vector algebra and analysis used in the text may be derived from the identity

$$\epsilon_{ijk} \epsilon_{ipq} = \delta_{jp} \delta_{kq} - \delta_{jq} \delta_{kp} \,. \qquad (1.3)$$

The simplest way to prove it is by direct substitution of specific values. As an application, let us prove in the component formalism the well-known identity

$$\nabla \times (\nabla \times \vec{A}) = \nabla \nabla \cdot \vec{A} - \nabla^2 \vec{A}. \qquad (1.4)$$

We start by writing the left-hand side in components

$$\left[\nabla \times (\nabla \times \vec{A})\right]_i = \epsilon_{ijk}\epsilon_{klm}\frac{\partial}{\partial x_j}\frac{\partial}{\partial x_l}A_m = \left(\delta_{il}\delta_{jm} - \delta_{im}\delta_{jl}\right)\frac{\partial}{\partial x_j}\frac{\partial}{\partial x_l}A_m =$$

$$= \frac{\partial}{\partial x_i}\frac{\partial}{\partial x_l}A_l - \frac{\partial}{\partial x_l}\frac{\partial}{\partial x_l}A_i = \left[\nabla\nabla\cdot\vec{A} - \nabla^2\vec{A}\right]_i,$$

where in the second equality we used Eq. (1.3) and the relation $\epsilon_{ijk} = \epsilon_{kij}$ following from the above definition of Levi–Civita symbol. Other well-known identities of vector analysis,

$$\nabla\cdot(\nabla\times\vec{A}) = \epsilon_{jki}\frac{\partial}{\partial x_j}\frac{\partial}{\partial x_k}A_i = 0 \tag{1.5}$$

and

$$[\nabla\times\nabla\varphi]_i = \epsilon_{ijk}\frac{\partial}{\partial x_j}\frac{\partial}{\partial x_k}\varphi = 0, \tag{1.6}$$

follow from the observation that the combination of the Levi–Civita symbol, which exhibits antisymmetry under the exchange of j and k, and the symmetric nature of partial derivatives $\frac{\partial}{\partial x_j}$ and $\frac{\partial}{\partial x_k}$ in relation to the interchange of j and k vanish.

The Gauss theorem of vector analysis reads in the component notation

$$\int d^3\vec{r}\,\frac{\partial f(\vec{r})}{\partial x_k} = \oint dS_k\,f(\vec{r})\,. \tag{1.7}$$

1.4 Dirac δ-function

In the next chapter, we will need to operate with the Dirac δ-function. From a physical point of view, it is the best to regard the δ-function as a limit of functions, for instance:

$$\delta(x) = \lim_{\varepsilon\to 0+}\frac{1}{\pi}\frac{\varepsilon}{x^2 + \varepsilon^2}\,, \tag{1.8}$$

that is,

$$\delta(x\neq 0) = 0\,, \qquad \delta(x = 0) = \infty\,.$$

From Eq. (1.8), we easily obtain the integral of the δ-function along the entire line[1]

$$\int_{-\infty}^{\infty} \delta(x)\,dx = \frac{1}{\pi}\lim_{\varepsilon\to 0+}\int_{-\infty}^{\infty}\frac{\varepsilon}{x^2+\varepsilon^2}dx = \lim_{\varepsilon\to 0+} 1 = 1\,, \tag{1.9}$$

and also with the Fourier transform

$$\begin{aligned}
\delta(x) &= \lim_{\varepsilon\to 0+}\frac{1}{\pi}\Im\frac{1}{x-i\varepsilon} = \lim_{\varepsilon\to 0+}\frac{1}{2i\pi}\left[\frac{1}{x-i\varepsilon}-\frac{1}{x+i\varepsilon}\right]\\
&= \lim_{\varepsilon\to 0+}\frac{1}{2\pi}\left[\int_0^{\infty} e^{-ik(x-i\varepsilon)}\,dk + \int_0^{\infty} e^{ik(x+i\varepsilon)}\,dk\right]\\
&= \lim_{\varepsilon\to 0+}\frac{1}{2\pi}\int_{-\infty}^{\infty} e^{-ikx-|k|\varepsilon}\,dk\,.
\end{aligned} \tag{1.10}$$

The integral of a product of the δ-function and an ordinary function equals

$$\int_{-\infty}^{\infty} f(x)\delta(x-a)\,dx = f(a)\,. \tag{1.11}$$

We can easily obtain this equation from the Taylor expansion of the function $f(x)$

$$f(x) = f(a) + (x-a)f'(a) + \frac{1}{2!}(x-a)^2 f''(a) + \dots\,,$$

from the normalization condition (1.9), and from the equation

$$\int x^n\delta(x)dx = 0\,, \quad n > 0\,.$$

The three-dimensional δ-function is the product of three one-dimensional δ-functions:

$$\delta(\vec{r}-\vec{r}') = \delta(x-x')\delta(y-y')\delta(z-z')\,. \tag{1.12}$$

Generalization of the above relations to three dimensions is straightforward; in particular, generalization of Eq. (1.11),

$$\int d^3\vec{r}\, f(\vec{r})\delta(\vec{r}-\vec{a}) = f(\vec{a})\,, \tag{1.13}$$

holds true.

---

[1] When solving such problems, we start with the definition of the δ-function (1.8) and leave the limit $\varepsilon \to 0+$ to the very end of our calculation.

1.5 Spherical Coordinates

We often make transition between the Cartesian and spherical coordinates

$$\vec{r} = r\vec{n}, \quad \vec{n}\cdot\vec{n} = 1, \quad d^3\vec{r} = dr\,r^2d\Omega, \quad d\vec{S} = r^2d\Omega\vec{n}, \tag{1.14}$$

where

$$\vec{n} = (\sin\vartheta\cos\varphi, \sin\vartheta\sin\varphi, \cos\vartheta) \tag{1.15}$$

and

$$\int d\Omega f(\vec{n}) = \int_0^\pi d\vartheta \int_0^{2\pi} d\varphi \sin\vartheta f(\vec{n}). \tag{1.16}$$

For differentiation of the functions depending only on the distance, we have

$$\nabla f(r) = \vec{n}\frac{df(r)}{dr}, \quad \nabla^2 f(r) = \frac{d^2 f(r)}{dr^2} + \frac{2}{r}\frac{df(r)}{dr}. \tag{1.17}$$

Chapter 2
Classical Electrodynamics

Starting from Maxwell and Newton equations, we develop here the theory of emission, dispersion, absorption, and scattering of light by matter based on the oscillator model of the matter. This is the standard material that can be found in many places, see, e.g., [5, 6]. For our purposes, the most important is the final Sect. 2.10 where we introduce the notion of oscillator strengths and Thomas–Kuhn–Reiche sum rule.

2.1 Maxwell and Newton Equations

Classical electrodynamics is a mathematical theory of the electromagnetic interaction between charged massive points. It consists of two sets of equations: the Maxwell equations and the Newton equation of motion with the Lorentz force. The Maxwell equations,

$$\nabla \times \vec{E} + \frac{\partial \vec{B}}{\partial t} = 0, \qquad \nabla \cdot \vec{B} = 0, \tag{2.1}$$

$$\nabla \times \vec{B} - \frac{\partial \vec{E}}{\partial t} = \vec{j}, \qquad \nabla \cdot \vec{E} = \rho, \tag{2.2}$$

describe how charge densities and currents determine the electric and magnetic fields. The charge densities and currents are supposed to be created by N charged massive points

$$\rho(\vec{r}, t) = \sum_{n=1}^{N} e_n \delta(\vec{r} - \vec{q}^{\,(n)}(t)), \qquad \vec{j}(\vec{r}, t) = \sum_{n=1}^{N} e_n \dot{\vec{q}}^{\,(n)}(t) \delta(\vec{r} - \vec{q}^{\,(n)}(t)). \tag{2.3}$$

© The Author(s), under exclusive license to Springer Nature Switzerland AG 2023
J. Zamastil, *Understanding the Path from Classical to Quantum Mechanics*,
SpringerBriefs in Physics, https://doi.org/10.1007/978-3-031-37373-2_2

The Newton equation of motion with the Lorentz force describes how the electric and magnetic fields act on massive points

$$m_n \ddot{\vec{q}}^{\,(n)} = e_n(\vec{E} + \dot{\vec{q}}^{\,(n)} \times \vec{B})\,. \tag{2.4}$$

Clearly, e_n and m_n are the charge and mass of the n-th massive point, respectively.

2.2 The Decomposition of the Electromagnetic Field

Applying the divergence operation on both sides of the first Eq. (2.2), we get, by virtue of Eq. (1.5) and the second Eq. (2.2), the law of charge conservation

$$-\frac{\partial \rho}{\partial t} = \nabla \cdot \vec{j}\,. \tag{2.5}$$

Substituting for charge densities and currents from Eq. (2.3), we get

$$-\frac{\partial \rho}{\partial t} = \sum_{n=1}^{N} e_n \frac{\partial \delta(\vec{r} - \vec{q}^{\,(n)}(t))}{\partial x_i} \dot{q}_i^{\,(n)}(t)$$

and

$$\nabla \cdot \vec{j} = \sum_{n=1}^{N} e_n \dot{q}_i^{\,(n)}(t) \frac{\partial \delta(\vec{r} - \vec{q}^{\,(n)}(t))}{\partial x_i}\,.$$

Clearly, for charge densities and currents defined by Eq. (2.3), the law of charge conservation, Eq. (2.5), holds true.

For further consideration, it is advantageous to decompose the electric intensity $\vec{E}$ and the charge current density $\vec{j}$ into transversal and longitudinal parts,

$$\vec{E} = \vec{E}_\perp + \vec{E}_{||}\,, \qquad \vec{j} = \vec{j}_\perp + \vec{j}_{||}\,, \tag{2.6}$$

where for a vector field $\vec{V}$ the transversal, $\vec{V}_\perp$, and the longitudinal, $\vec{V}_{||}$, parts are divergenceless and rotationless, respectively,

$$\nabla \cdot \vec{V}_\perp = 0\,, \qquad \nabla \times \vec{V}_{||} = 0\,. \tag{2.7}$$

Given a vector field $\vec{V}$, the transversal and longitudinal parts of the vector field read

$$(V_\perp)_i = \left(\delta_{ij} - \frac{\frac{\partial}{\partial x_i}\frac{\partial}{\partial x_j}}{\nabla^2} \right) V_j\,, \qquad (V_{||})_i = \frac{\frac{\partial}{\partial x_i}\frac{\partial}{\partial x_j}}{\nabla^2} V_j\,, \qquad V_i = (V_\perp)_i + (V_{||})_i\,.$$

The reader does not have to worry about the meaning of the operation $\frac{1}{\nabla^2}$. It suffices to think of it as an inverse operator to the operator ∇^2, meaning that $\nabla^2 \frac{1}{\nabla^2} = \frac{1}{\nabla^2} \nabla^2 = 1$, where 1 denotes the identity operator that does nothing. Clearly,

$$\nabla \cdot \vec{V}_{\perp} = \frac{\partial (V_{\perp})_i}{\partial x_i} = \left(\frac{\partial}{\partial x_j} - \frac{\frac{\partial}{\partial x_i} \frac{\partial}{\partial x_i} \frac{\partial}{\partial x_j}}{\nabla^2} \right) V_j = 0,$$

where the definition of Laplace operator, Eq. (1.2), has been used, and cf. Eq. (1.6),

$$\left(\nabla \times \vec{V}_{||} \right)_i = \epsilon_{ijk} \frac{\partial (V_{||})_k}{\partial x_j} = \epsilon_{ijk} \frac{\frac{\partial}{\partial x_j} \frac{\partial}{\partial x_k} \frac{\partial}{\partial x_l}}{\nabla^2} V_l = 0.$$

Inserting the decomposition of the electric intensity and charge current density, Eq. (2.6), into Maxwell equations, Eqs. (2.1) and (2.2), we obtain

$$\nabla \times \vec{E}_{\perp} + \frac{\partial \vec{B}}{\partial t} = 0, \qquad \nabla \cdot \vec{B} = 0, \tag{2.8}$$

$$\nabla \times \vec{B} - \frac{\partial \vec{E}_{\perp}}{\partial t} - \frac{\partial \vec{E}_{||}}{\partial t} = \vec{j}_{\perp} + \vec{j}_{||}, \qquad \nabla \cdot \vec{E}_{||} = \rho. \tag{2.9}$$

Applying the operation divergence on the first Eq. (2.9), we obtain

$$-\frac{\partial \nabla \cdot \vec{E}_{||}}{\partial t} = \nabla \cdot \vec{j}_{||}.$$

Substituting into the last equation for the divergence of the longitudinal part of the electric intensity from the second Eq. (2.9), we obtain

$$-\frac{\partial \rho}{\partial t} = \nabla \cdot \vec{j}_{||}.$$

This is the charge conservation equation, Eq. (2.5). In view of the last two equations, we set

$$-\frac{\partial \vec{E}_{||}}{\partial t} = \vec{j}_{||}.$$

By inserting the last equality into the first Eq. (2.9), we obtain

$$\nabla \times \vec{B} - \frac{\partial \vec{E}_{\perp}}{\partial t} = \vec{j}_{\perp}. \tag{2.10}$$

Differentiating the last equation with respect to time and substituting for $\frac{\partial \vec{B}}{\partial t}$ from the first Eq. (2.8), we obtain

$$\nabla^2 \vec{E}_\perp - \frac{\partial^2 \vec{E}_\perp}{\partial t^2} = \frac{\partial \vec{j}_\perp}{\partial t}, \tag{2.11}$$

where the identity Eq. (1.4) and definition (2.7) have been used.

The physical meaning of the decomposition (2.6) should be clear now. The longitudinal part of the electric field satisfies equations for the electrostatic field

$$\nabla \times \vec{E}_{\|} = 0, \qquad \nabla \cdot \vec{E}_{\|} = \rho,$$

and hence describes the electrostatic interaction between charges. Since the rotation of the gradient of a scalar potential is identically zero, $\nabla \times \nabla \varphi = 0$, see Eq. (1.6), one can write the longitudinal part of the electric field as a gradient of a scalar potential,

$$\vec{E}_{\|} = -\nabla \varphi, \qquad -\nabla^2 \varphi(\vec{r}, t) = \rho(\vec{r}, t), \tag{2.12}$$

where the second equation follows from inserting the first equation into the second Eq. (2.9). The transversal part of the field satisfies the wave equation (2.11) and describes the dynamical part of the field.

2.3 Conservation of Energy for a System of N Particles and Electromagnetic Field

To formulate the energy conservation, we take the dot product of Maxwell Eq. (2.1), $\nabla \times \vec{E} + \frac{\partial \vec{B}}{\partial t} = 0$, with $\vec{B}$ and the dot product of Maxwell Eq. (2.2), $\nabla \times \vec{B} - \frac{\partial \vec{E}}{\partial t} = \vec{j}$, with $\vec{E}$. We subtract the two equations and get

$$\frac{\partial}{\partial t} \frac{1}{2} \left[\vec{E} \cdot \vec{E} + \vec{B} \cdot \vec{B} \right] + \nabla \cdot (\vec{E} \times \vec{B}) = -\vec{j} \cdot \vec{E}.$$

Substituting for the charge current, Eq. (2.3), $\vec{j}(\vec{r}, t) = \sum_{n=1}^{N} e_n \dot{\vec{q}}^{(n)}(t) \delta(\vec{r} - \vec{q}^{(n)}(t))$, and integrating over the volume enclosing the N charged particles, we get by means of the Gauss theorem, Eq. (1.7), and the property of the δ-function, Eq. (1.13),

$$\frac{d}{dt} \int d^3 r \frac{1}{2} \left[\vec{E} \cdot \vec{E} + \vec{B} \cdot \vec{B} \right] + \oint d\vec{s} \cdot (\vec{E} \times \vec{B}) = -\sum_{n=1}^{N} e_n \dot{\vec{q}}^{(n)} \cdot \vec{E}.$$

Further, we take the dot product of Newton Eq. (2.4), $m_n \ddot{\vec{q}}^{(n)} = e_n(\vec{E} + \dot{\vec{q}}^{(n)} \times \vec{B})$, with $\dot{\vec{q}}^{(n)}$ to get

$$\frac{d}{dt}\left(\frac{1}{2}m_n \dot{\vec{q}}^{(n)} \cdot \dot{\vec{q}}^{(n)}\right) = e_n \dot{\vec{q}}^{(n)} \cdot \vec{E}.$$

Adding now the last two equations, we finally get energy conservation for the system of the N charged particles and electromagnetic field,

$$-\frac{d\mathcal{E}}{dt} = \mathcal{P}, \quad \mathcal{E} = \frac{1}{2}\int d^3\vec{r}\left(\vec{E} \cdot \vec{E} + \vec{B} \cdot \vec{B}\right) + \frac{1}{2}\sum_{j=1}^{N} m_j \dot{\vec{q}}^{(j)} \cdot \dot{\vec{q}}^{(j)},$$

$$\mathcal{P} = \oint d\vec{S} \cdot (\vec{E} \times \vec{B}). \tag{2.13}$$

The decrease of energy inside a volume equals the energy flux, the outgoing power, over the surface of the volume. For the following considerations, it is important to note that we are interested in the outgoing power that does not vanish as the volume extends over the whole space. The differential element of the surface changes like r^2, where r is the radius of the volume, see Eq. (1.14). Thus, only radiative part of the electromagnetic fields $\vec{E}$ and $\vec{B}$, one that changes as r^{-1}, contributes to the outgoing power at infinity.

2.4 Solution of Maxwell Equations

Let us start with the solution of the scalar Eq. (2.12). This equation is solved by the method of Green functions. We first find the solution of the equation

$$-\nabla^2 G(\vec{r}, \vec{r}') = \delta(\vec{r} - \vec{r}'). \tag{2.14}$$

The solution of Eq. (2.12) then reads

$$\varphi(\vec{r}) = \int d^3\vec{r}' G(\vec{r}, \vec{r}')\rho(\vec{r}'), \tag{2.15}$$

as can be verified by direct substitution

$$-\nabla^2\varphi(\vec{r}) = -\int d^3\vec{r}' \nabla^2 G(\vec{r}, \vec{r}')\rho(\vec{r}') = \int d^3\vec{r}' \delta(\vec{r} - \vec{r}')\rho(\vec{r}') = \rho(\vec{r}),$$

where in the last equality we use the property of the δ-function, Eq. (1.13). The solution of Eq. (2.14) is greatly simplified if we put the origin of the coordinate system in $\vec{r}' = \vec{0}$, since the problem then becomes spherically symmetric.

Integrating Eq. (2.14) over the whole space, using Gauss theorem, Eq. (1.7), and introducing spherical coordinates, Eq. (1.14), we get

$$-r^2 \int d\Omega \vec{n} \cdot \nabla G(r) = 1 \quad \Rightarrow \quad G(r) = \frac{1}{4\pi r}. \tag{2.16}$$

The implication follows from Eq. (1.17). Substituting this solution into Eq. (2.15), we obtain

$$\varphi(\vec{r}, t) = \frac{1}{4\pi} \int d^3 \vec{r}' \frac{\rho(\vec{r}', t)}{|\vec{r} - \vec{r}'|}. \tag{2.17}$$

Next, let us solve the equation

$$\left(\frac{\partial^2}{\partial t^2} - \nabla^2 \right) \vec{E}(\vec{r}, t) = -\frac{\partial \vec{j}(\vec{r}, t)}{\partial t}. \tag{2.18}$$

We solve it by the method of the time-dependent Green functions

$$\left(\frac{\partial^2}{\partial t^2} - \nabla^2 \right) G(\vec{r}, t, \vec{r}', t') = \delta(\vec{r} - \vec{r}')\delta(t - t'). \tag{2.19}$$

The solution of Eq. (2.18) is then

$$\vec{E}(\vec{r}, t) = -\int d^3 \vec{r}' \int dt' G(\vec{r}, t, \vec{r}', t') \frac{\partial \vec{j}(\vec{r}', t')}{\partial t'}, \tag{2.20}$$

as can be verified by direct substitution. Equation (2.19) is solved by the method of Fourier transformation, cf. Eq. (1.10),

$$G(\vec{r}, t, \vec{r}', t') = \frac{1}{2\pi} \int_{-\infty}^{\infty} d\omega G(\vec{r}, \vec{r}', \omega) e^{i\omega(t-t')}, \quad \delta(t - t') = \frac{1}{2\pi} \int_{-\infty}^{\infty} d\omega e^{i\omega(t-t')}; \tag{2.21}$$

inserting these expressions into Eq. (2.19), we obtain

$$-\left(\omega^2 + \nabla^2 \right) G(\vec{r}, \vec{r}', \omega) = \delta(\vec{r} - \vec{r}'). \tag{2.22}$$

Again we put the origin of the coordinate system in $\vec{r}' = \vec{0}$. Thus, for $\vec{r} \neq \vec{0}$, we are to solve

$$\left(\omega^2 + \nabla^2 \right) G(r, \omega) = 0 \quad \Rightarrow \quad \left(\omega^2 + \frac{d^2}{dr^2} + \frac{2}{r} \frac{d}{dr} \right) G(r, \omega) = 0$$

$$\Rightarrow \quad G(r, \omega) = \frac{1}{4\pi} \frac{e^{\pm i\omega r}}{r}.$$

The first implication follows from Eq. (1.17), and the second implication follows from the requirement that for $\omega \to 0$ the solution has to approach the time-independent solution, Eq. (2.16). Substituting the Fourier transform back into Eq. (2.21), we obtain for the time-dependent Green function, by virtue of Eq. (1.10),

$$G(\vec{r}, t, \vec{r}\,', t') = \frac{1}{2\pi} \int_{-\infty}^{\infty} d\omega \frac{1}{4\pi} \frac{e^{\pm i\omega|\vec{r} - \vec{r}\,'|}}{|\vec{r} - \vec{r}\,'|} e^{i\omega(t - t')} = \frac{1}{4\pi} \frac{\delta(t - t' \pm |\vec{r} - \vec{r}\,'|)}{|\vec{r} - \vec{r}\,'|}.$$

The two signs in δ-function correspond to the advanced or retarded solutions. The field at a specific time t and position $\vec{r}$ is influenced by variations in the current density over time at the location $\vec{r}\,'$. These variations can occur either at a later time $t' = t + |\vec{r} - \vec{r}\,'|$ or at an earlier time $t' = t - |\vec{r} - \vec{r}\,'|$. Choosing the retarded solution and substituting it into Eq. (2.20), we find the solution of Eq. (2.18) reads

$$\vec{E}(\vec{r}, t) = -\frac{1}{4\pi} \int d^3\vec{r}\,' \frac{\frac{\partial \vec{j}(\vec{r}\,', t' = t - |\vec{r} - \vec{r}\,'|)}{\partial t'}}{|\vec{r} - \vec{r}\,'|}.$$

The solution of the vector Maxwell Eq. (2.11) is then obtained by multiplying the last equation by the projector $\left(1 - \frac{\nabla\nabla\cdot}{\nabla^2}\right)$ to guarantee that solution is divergenceless

$$(E_\perp)_k(\vec{r}, t) = -\left(\delta_{kl} - \frac{\frac{\partial}{\partial x_k} \frac{\partial}{\partial x_l}}{\nabla^2}\right) \frac{1}{4\pi} \int d^3\vec{r}\,' \frac{\frac{\partial j_l(\vec{r}\,', t' = t - |\vec{r} - \vec{r}\,'|)}{\partial t'}}{|\vec{r} - \vec{r}\,'|}. \tag{2.23}$$

2.5 Asymptotic Form of Solution

We are interested in the emission of radiation, the outgoing power at infinity, Eq. (2.13). Thus, we only need the asymptotic form of the above solution valid for large r. To obtain this form, we make replacements

$$\frac{1}{|\vec{r} - \vec{r}\,'|} \simeq \frac{1}{r}, \quad |\vec{r} - \vec{r}\,'| \simeq r - \vec{n} \cdot \vec{r}\,'$$

in the integrands in Eqs. (2.17) and (2.23). Further, for differentiation with respect to space coordinates, we can write for large r

$$\frac{\partial}{\partial x_k} \frac{1}{r} \int d^3\vec{r}\,' j_l\left(\vec{r}\,', t' = t - r + \vec{n} \cdot \vec{r}\,'\right) \simeq \frac{1}{r} \int d^3\vec{r}\,' \frac{\partial j_l}{\partial t'} \frac{\partial t'}{\partial x_k} \simeq -\frac{n_k}{r} \int d^3\vec{r}\,' \frac{\partial j_l}{\partial t'}. \tag{2.24}$$

The neglected terms decay like r^{-2} for large r and hence do not contribute to the outgoing power at infinity. Thus, the asymptotic form of the solutions (2.17) and (2.23) valid for large r reads

$$(E_\perp)_k(\vec{r}, t) \simeq -(\delta_{kl} - n_k n_l) \frac{1}{4\pi r} \int d^3\vec{r}\,' \frac{\partial j_l \left(\vec{r}\,', t' = t - r + \vec{r}\,' \cdot \vec{n}\right)}{\partial t'}, \qquad (2.25)$$

$$\varphi(\vec{r}, t) \simeq \frac{Q}{4\pi r}, \qquad Q = \int d^3\vec{r}\,' \rho(\vec{r}\,', t),$$

where Q is the total charge. It follows from Eq. (2.12) that the longitudinal part of the electric field, $\vec{E}_\|$, decays like r^{-2} for large r and does not contribute to the outgoing power at infinity. Thus, for large r, the electric intensity is given only by its transversal part,

$$\vec{E} \simeq \vec{E}_\perp . \qquad (2.26)$$

Since $n_k (\delta_{kl} - n_k n_l) = 0$, we see that in the asymptotic region, the transversality condition, Eq. (2.7), is translated into the projection of the field into the plane perpendicular to the line of sight.

The asymptotic form of magnetic induction $\vec{B}$ is determined from the first Eq. (2.8), $\nabla \times \vec{E}_\perp + \frac{\partial \vec{B}}{\partial t} = 0$. As follows from foregoing considerations, the electric intensity varies with space for large r only through the retarded time t'; thus, cf. Eq. (2.24),

$$\nabla \times \vec{E}_\perp \simeq -\vec{n} \times \frac{\partial \vec{E}}{\partial t'} .$$

Since $\frac{\partial \vec{B}}{\partial t} = \frac{\partial \vec{B}}{\partial t'}$, for large r, one can write

$$\vec{B} \simeq \vec{n} \times \vec{E}_\perp . \qquad (2.27)$$

Thus, for large distances from any form of charge distribution, the magnetic induction is perpendicular to both the electric intensity and the line of sight.

2.6   Emission of Radiation

Returning now to the formula (2.13) for the outgoing power at infinity, we successively obtain

$$\mathcal{P} = \oint d\vec{S} \cdot (\vec{E} \times \vec{B}) \simeq r^2 \int d\Omega \vec{n} \cdot \left(\vec{E}_\perp \times (\vec{n} \times \vec{E}_\perp)\right)$$

$$\simeq r^2 \int d\Omega \left[\vec{E}_\perp \cdot \vec{E}_\perp - (\vec{n} \cdot \vec{E}_\perp)^2\right], \qquad (2.28)$$

where in the second equality we substituted Eqs. (1.14), (2.26), and (2.27), and in the third equality, we used the identity (1.3). Substituting now in the last equation for $\vec{E}_\perp$ from Eq. (2.25), we obtain the final exact formula for the radiated power at infinity

$$
\mathcal{P} = \int d\Omega \frac{\delta_{kl} - n_k n_l}{(4\pi)^2} \int d^3\vec{r}\,' \frac{\partial j_k(\vec{r}\,', t' = t - r + \vec{n}\cdot\vec{r}\,')}{\partial t'}
$$

$$
\times \int d^3\vec{r}\,'' \frac{\partial j_l(\vec{r}\,'', t'' = t - r + \vec{n}\cdot\vec{r}\,'')}{\partial t''} . \tag{2.29}
$$

For our purposes, it suffices to neglect retardation,

$$
j_l(\vec{r}\,', t' = t - r + \vec{n}\cdot\vec{r}\,') \simeq j_l(\vec{r}\,', t' = t) .
$$

Further, substituting for charge current density from Eq. (2.3), we obtain

$$
\mathcal{P} \simeq \int d\Omega \frac{\delta_{kl} - n_k n_l}{(4\pi)^2} \sum_{n=1}^{N} e_n \dddot{q}_k^{(n)}(t) \sum_{m=1}^{N} e_m \dddot{q}_l^{(m)}(t)
$$

$$
= \frac{2}{3}\frac{1}{4\pi} \left(\sum_{n=1}^{N} e_n \dddot{\vec{q}}^{(n)}(t) \right) \cdot \left(\sum_{m=1}^{N} e_m \dddot{\vec{q}}^{(m)}(t) \right) . \tag{2.30}
$$

In the last equality, we used[1]

$$
\int d\Omega\, n_k n_l = \frac{4\pi}{3}\delta_{kl} . \tag{2.31}
$$

For one charged particle, we get for the radiated power the Larmor formula

$$
\mathcal{P} = \frac{2}{3}\alpha \dddot{\vec{q}} \cdot \dddot{\vec{q}} , \tag{2.32}
$$

where we substituted from Eq. (1.1).

---

[1] This identity can be proven either by the direct substitution of Eqs. (1.15) and (1.16) into Eq. (2.31) or from symmetry considerations: on the right, we need quantity with two indices, and the only available is Kronecker delta. The numerical factor follows from setting $l = k$ and using Einstein summation convention; then $n_k n_k = 1$, $\delta_{kk} = 3$ and $\int d\Omega = 4\pi$.

2.7   Representation of Radiation Damping as Friction

It is seen that once the motion of the charged particle is not uniform, the particle loses its energy by emitting the electromagnetic radiation. This is usually referred to as radiation damping. If we leave out the energy stored in the electromagnetic field, the term $\frac{1}{2} \int d^3 r \left(\vec{E} \cdot \vec{E} + \vec{B} \cdot \vec{B} \right)$ in Eq. (2.13), this radiation damping can be represented as an effective friction force. For the charged particle moving in the potential V, we have

$$m \ddot{\vec{q}} = -\vec{F}_{\text{fr}} - \nabla V . \tag{2.33}$$

By multiplying this equation by velocity, we obtain the law of energy conservation in the form

$$-\frac{d\mathcal{E}}{dt} = \vec{F}_{\text{fr}} \cdot \dot{\vec{q}} , \quad \mathcal{E} = \frac{m}{2} \dot{\vec{q}} \cdot \dot{\vec{q}} + V . \tag{2.34}$$

Now we identify the decay of the particle energy with time with the emitted power, Eq. (2.32),

$$\vec{F}_{\text{fr}} \cdot \dot{\vec{q}} = \frac{2}{3} \alpha \ddot{\vec{q}} \cdot \ddot{\vec{q}} . \tag{2.35}$$

For instance, if the particle moves in the parabolic potential

$$V = \frac{m \omega^2}{2} \vec{q} \cdot \vec{q} , \tag{2.36}$$

then according to the Newton equation of motion, Eq. (2.33), where we first neglect the effect of friction, the particle executes the oscillatory motion

$$\vec{q} = \vec{q}_0 \cos(\omega t + \varphi) , \tag{2.37}$$

where $\vec{q}_0$ and φ are the amplitude of the oscillations and initial phase, respectively. Then we get from Eq. (2.35)

$$\vec{F}_{\text{fr}} = m \Gamma \dot{\vec{q}} , \quad \Gamma = \frac{2}{3m} \alpha \omega^2 . \tag{2.38}$$

The parameter Γ determines the average rate of energy loss of the harmonic oscillator per one oscillation

$$\frac{\overline{\frac{d\mathcal{E}}{dt}}}{\mathcal{E}} = \frac{\frac{2}{3} \alpha \overline{\ddot{\vec{q}} \cdot \ddot{\vec{q}}}}{\frac{m}{2} \left(\dot{\vec{q}} \cdot \dot{\vec{q}} + \omega^2 \vec{q} \cdot \vec{q} \right)} = \Gamma , \tag{2.39}$$

where in the first equality we substituted for the outgoing power from Eq. (2.32) and for the energy from Eqs. (2.34) and (2.36). In the second equality, we substituted from Eq. (2.37). The overline denotes the average over one oscillation period

$$\overline{Q} = \frac{1}{T} \int_0^T dt\, Q(t)\,, \quad \omega = \frac{2\pi}{T}\,. \tag{2.40}$$

For instance, in Eq. (2.39), we used the equality

$$\overline{\ddot{\vec{q}} \cdot \ddot{\vec{q}}} = \vec{q}_0 \cdot \vec{q}_0 \omega^2 \overline{\cos^2(\omega t + \varphi)} = \frac{\omega^2}{2} \vec{q}_0 \cdot \vec{q}_0\,.$$

In view of Eq. (2.39), the inverse of Γ represents the oscillator lifetime

$$\tau = \frac{1}{\Gamma}\,. \tag{2.41}$$

2.8 Dispersion and Absorption of Radiation

Consider an electron, bound to the atom by elastic force, to be under influence of the long-wavelength monochromatic electromagnetic (EM) radiation. Neglecting the action of the magnetic field (the radiation is long wavelength), the equation of electron motion is then, see Eqs. (2.4), (2.33), and (2.36),

$$m\ddot{\vec{q}} = -m\omega^2 \vec{q} - m\Gamma \dot{\vec{q}} + e\vec{E}_\perp(t)\,, \quad \vec{E}_\perp(t) = \vec{E}_0 \cos(\omega_e t)\,, \tag{2.42}$$

where $\vec{E}_0$ is the amplitude and polarization of the EM wave and ω_e its circular frequency. We search for the particular solution of this equation in the form

$$\vec{q}(t) = \Re \vec{q}_0 e^{i\omega_e t}\,; \tag{2.43}$$

by substituting this expression into Eq. (2.42), we obtain

$$\vec{q}_0 = \frac{e}{m} \vec{E}_0 \frac{1}{\omega^2 - \omega_e^2 + i\Gamma \omega_e}\,. \tag{2.44}$$

Substituting back this solution into Eq. (2.43), we obtain

$$\vec{q}(t) = \frac{e}{m} \vec{E}_0 \frac{(\omega^2 - \omega_e^2)\cos(\omega_e t) + \Gamma \omega_e \sin(\omega_e t)}{(\omega^2 - \omega_e^2)^2 + (\Gamma \omega_e)^2}$$

$$= \frac{e}{m} \frac{(\omega^2 - \omega_e^2)}{(\omega^2 - \omega_e^2)^2 + (\Gamma \omega_e)^2} \vec{E}_\perp(t) - \frac{e}{m} \frac{\Gamma}{(\omega^2 - \omega_e^2)^2 + (\Gamma \omega_e)^2} \frac{\partial \vec{E}_\perp(t)}{\partial t}\,. \tag{2.45}$$

Integrating the charge current density, Eq. (2.3), over a small volume V, we obtain

$$\int d^3\vec{r}\,\vec{j} = \sum_{n=1}^{N} e_n \dot{\vec{q}}^{\,(n)} \simeq N e \dot{\vec{q}} \,,$$

where N is the number of atoms in the volume V. In the last equality, we assumed that the field interacting with all the N atoms within the volume V is the same. Thus, for the charge current density, we write

$$\vec{j}(\vec{r}, t) = \sum_{n=1}^{N} e_n \dot{\vec{q}}^{\,(n)}(t)\delta(\vec{r} - \vec{q}^{\,(n)}(t)) \simeq N e \dot{\vec{q}} \,,$$

where henceforth N stands for the number of atoms per unit volume. Substituting into the last equation from Eq. (2.45), we get

$$\vec{j}(\vec{r}, t) = a\frac{\partial \vec{E}_\perp(\vec{r}, t)}{\partial t} - b\frac{\partial^2 \vec{E}_\perp(\vec{r}, t)}{\partial t^2} \,, \tag{2.46}$$

where a and b denote the dispersive and absorptive parts of the polarizability, respectively,

$$a = N\frac{e^2}{m}\frac{(\omega^2 - \omega_e^2)}{(\omega^2 - \omega_e^2)^2 + (\Gamma\omega_e)^2} \,, \qquad b = N\frac{e^2}{m}\frac{\Gamma}{(\omega^2 - \omega_e^2)^2 + (\Gamma\omega_e)^2} \,. \tag{2.47}$$

Note that since $\nabla \cdot \vec{E}_\perp = 0$, the charge density given by Eq. (2.46) is divergenceless, $\nabla \cdot \vec{j} = 0$, so $\vec{j} = \vec{j}_\perp$. To determine propagation of the EM wave in the matter, we substitute Eq. (2.46) into the Maxwell Eq. (2.11):

$$\left(\frac{\partial^2}{\partial t^2} - \nabla^2\right)\vec{E}_\perp(\vec{r}, t) = -a\frac{\partial^2 \vec{E}_\perp(\vec{r}, t)}{\partial t^2} + b\frac{\partial^3 \vec{E}_\perp(\vec{r}, t)}{\partial t^3} \,. \tag{2.48}$$

We search for the solution of this equation in the form, cf. Eq. (2.42),

$$\vec{E}_\perp = \Re\vec{E}_0 e^{-i(\vec{k}_e \cdot \vec{r} - \omega_e t)} \,, \qquad \vec{k}_e \cdot \vec{E}_0 = 0 \,, \qquad \vec{k}_e = \vec{e}_z k_e \,, \tag{2.49}$$

where the second equation follows from the transversality condition, Eq. (2.7), and the third equation is the suitable choice of the coordinate system. We choose the z-axis in the direction of the propagation of EM wave, and $\vec{e}_z$ is the unit vector pointing along the z-axis. Since the atom size is of the order 10^{-10} meters, while the wavelength of the visible light is of the order 10^{-7} meters, we can neglect the space dependence of EM wave inside the atom, $\vec{E}_\perp \simeq \vec{E}_0 \cos(\omega_e t)$. Hence, the assumed form of solution, Eq. (2.49), is consistent with Eq. (2.42). Substitution of

Eq. (2.49) into Eq. (2.48) then yields

$$k_e^2 = (1 + a)\omega_e^2 - ib\omega_e^3 \quad \Rightarrow \quad k_e \simeq \omega_e \left[1 + \frac{a}{2} - i\frac{b\omega_e}{2} \right]. \tag{2.50}$$

Inserting this solution back into Eq. (2.49), we get

$$\vec{E}_\perp = \vec{E}_0 \cos(\phi) e^{-\frac{b}{2}\omega_e^2 z}, \quad \phi = \omega_e \left[\left(1 + \frac{a}{2} \right) z - t \right]. \tag{2.51}$$

Substituting the last equation into the first Eq. (2.8), $\nabla \times \vec{E}_\perp + \frac{\partial \vec{B}}{\partial t} = 0$, we get for the magnetic induction

$$\vec{B} = \vec{e}_z \times \vec{E}_0 e^{-\frac{b}{2}\omega_e^2 z} \left[\left(1 + \frac{a}{2} \right) \cos(\phi) - \frac{b}{2}\omega_e \sin(\phi) \right]. \tag{2.52}$$

Thus the magnetic induction is perpendicular to both the electric intensity and the direction of the propagation of the EM wave.

For the power transferred by the EM wave, we get from Eq. (2.13) $\mathcal{P} = \oint d\vec{S} \cdot (\vec{E} \times \vec{B})$, where $\vec{S} = S_z \vec{e}_z$, and Eqs. (2.49) and (2.52)

$$\mathcal{P} = \oint dS_z E_0^2 e^{-b\omega_e^2 z} \left[\left(1 + \frac{a}{2} \right) \cos^2(\phi) - \frac{b}{2}\omega_e \cos(\phi)\sin(\phi) \right]. \tag{2.53}$$

For the averaged transferred power per one oscillation per unit area, we get, cf. Eq. (2.40),

$$\frac{\overline{d\mathcal{P}}}{dS_z} \simeq \frac{E_0^2}{2}(1 + a)e^{-b\omega_e^2 z}. \tag{2.54}$$

The quantity on the left-hand side of the last equation is usually called the intensity of the EM wave. The last equation then has the form of the Lambert–Beer law

$$I = I_0 e^{-\kappa z}, \quad I_0 = \frac{E_0^2}{2}(1 + a), \quad \kappa = b\omega_e^2. \tag{2.55}$$

For the phase velocity of the propagating EM wave, the refractive index, we get from Eq. (2.51)

$$n = \frac{dz}{dt} \simeq 1 - \frac{a}{2}. \tag{2.56}$$

The refractive index is responsible for the refraction of light between, say, air and water, see, e.g., [5].

2.9   Scattering of Radiation

Let us now consider scattering of monochromatic radiation,

$$\vec{E}_\perp = \vec{E}_0 \cos \{\omega_e [z - t]\} , \tag{2.57}$$

on an atom. Inserting again this equation into the first Eq. (2.8), we get for the magnetic induction, cf. Eq. (2.52),

$$\vec{B} = \vec{e}_z \times \vec{E}_\perp . \tag{2.58}$$

Hence, the incoming intensity of EM radiation, that is the incoming power per unit area, reads

$$\frac{d\mathcal{P}}{dS_z} = \vec{e}_z \cdot \vec{E}_\perp \times \vec{B} = \vec{E}_\perp \cdot \vec{E}_\perp , \tag{2.59}$$

where in the last equality we substituted for the magnetic induction from Eq. (2.58). Substituting Eq. (2.57) into Eq. (2.59), we get according to Eq. (2.40) for the incoming intensity averaged over one oscillation

$$\overline{\frac{d\mathcal{P}}{dS_z}} = \frac{E_0^2}{2} . \tag{2.60}$$

The incoming radiation is scattered by an atom; the power radiated by an atom is given by Eqs. (2.32) and (2.45)

$$P_{\text{scatt}} = \frac{2}{3}\alpha\ddot{\vec{q}} \cdot \ddot{\vec{q}} = \frac{8\pi}{3}\left(\frac{\alpha}{m}\right)^2 E_0^2 \omega_e^4 \left[\frac{(\omega^2 - \omega_e^2)\cos(\omega_e t) + \Gamma\omega_e \sin(\omega_e t)}{(\omega^2 - \omega_e^2)^2 + (\Gamma\omega_e)^2}\right]^2 . \tag{2.61}$$

For the scattered power averaged over one oscillation, we get from Eq. (2.40) and the last equation

$$\overline{P_{\text{scatt}}} = \frac{8\pi}{3}\left(\frac{\alpha}{m}\right)^2 \frac{E_0^2}{2} \frac{\omega_e^4}{(\omega^2 - \omega_e^2)^2 + (\Gamma\omega_e)^2} . \tag{2.62}$$

For the quantity of experimental interest, the cross-section of EM radiation on the atom, the ratio of the averaged scattered power, and averaged incoming power per unit area, we get from Eqs. (2.60) and (2.62)

$$\sigma = \frac{\overline{P_{\text{scatt}}}}{\overline{\frac{dP_{\text{in}}}{dS_z}}} = \frac{8\pi}{3}\left(\frac{\alpha}{m}\right)^2 \frac{\omega_e^4}{(\omega^2 - \omega_e^2)^2 + (\Gamma\omega_e)^2} . \tag{2.63}$$

There are two limiting cases of Eq. (2.63) that are of great interest. First, when the frequency of the incoming radiation ω_e is much lower than the atomic frequency ω, $\omega_e << \omega$, then

$$\sigma \simeq \frac{8\pi}{3} \left(\frac{\alpha}{m}\right)^2 \left(\frac{\omega_e}{\omega}\right)^4 . \tag{2.64}$$

This is called the Rayleigh scattering, and its dependence on the fourth power of the incident light explains why the sky is blue—the larger the frequency of the light, the greater is the scattering of the light on atoms at the atmosphere. Second, when the frequency of the incoming radiation ω_e is much higher than the atomic frequency ω, $\omega_e >> \omega$, then the cross-section approaches a constant value, the so-called Thomson scattering,

$$\sigma \simeq \frac{8\pi}{3} \left(\frac{\alpha}{m}\right)^2 . \tag{2.65}$$

2.10 Comment on Classical Theory

It turns out that the atom does not have one, but several characteristic frequencies. It has been firmly established in the first quarter of the twentieth century that if the formulae (2.43) and (2.44) for the induced electron radius vector are replaced by

$$\vec{q}(t) = \frac{e}{m} \vec{E}_0 \Re \sum_k \frac{f_k e^{i\omega_e t}}{\omega_k^2 - \omega_e^2 + i\Gamma_k \omega_e} , \tag{2.66}$$

where f_k are called oscillator strengths[2], the resulting formulae for the polarizability

$$a = N\frac{e^2}{m} \sum_k \frac{(\omega_k^2 - \omega_e^2) f_k}{(\omega_k^2 - \omega_e^2)^2 + (\Gamma_k \omega_e)^2} , \qquad b = N\frac{e^2}{m} \sum_k \frac{\Gamma_k f_k}{(\omega_k^2 - \omega_e^2)^2 + (\Gamma_k \omega_e)^2} , \tag{2.67}$$

---

[2] As we show later, see Eqs. (4.18) and (4.55), the oscillator strengths equal in modern notation

$$f_k = \frac{2}{3} m \omega_{k0} |\vec{q}_{0k}|^2 ,$$

where the circular frequency ω_{k0} is the difference between the energies of the k-th excited and ground stationary states, see Eq. (4.4), and $\vec{q}_{0k}$ is the matrix element of the coordinate between the ground and k-th excited states, see Eq. (5.7).

and the cross-section

$$
\sigma = \frac{8\pi}{3} \left(\frac{\alpha}{m}\right)^2 \omega_e^4 \left[\left(\sum_k f_k \frac{\omega_k^2 - \omega_e^2}{(\omega_k^2 - \omega_e^2)^2 + (\Gamma_k \omega_e)^2}\right)^2 \right.
$$

$$
\left. + \left(\sum_k f_k \frac{\Gamma_k \omega_e}{(\omega_k^2 - \omega_e^2)^2 + (\Gamma_k \omega_e)^2}\right)^2\right] \tag{2.68}
$$

are in agreement with experiment. At this level of development, the oscillator strengths f_k are determined empirically to fit the experimental data. The inverse lifetimes Γ_k are very small quantities, cf. Eqs. (1.1) and (2.38), and are of significance only in the case of resonant scattering, i.e., when the frequency of the incident light ω_e approaches one of the characteristic atomic frequencies ω_k. If this is not the case, i.e., there is no resonance, then the last two equations simplify to

$$
a \simeq N \frac{e^2}{m} \sum_k \frac{f_k}{\omega_k^2 - \omega_e^2}, \quad b \simeq 0, \tag{2.69}
$$

and

$$
\sigma \simeq \frac{8\pi}{3} \left(\frac{\alpha}{m}\right)^2 \omega_e^4 \left(\sum_k f_k \frac{1}{\omega_k^2 - \omega_e^2}\right)^2. \tag{2.70}
$$

If the frequency of the incident radiation is much larger than any of the atom's characteristic frequencies, the cross-section has to approach the Thomson value, Eq. (2.65). Hence, we get that

$$
\sum_k f_k = 1; \tag{2.71}
$$

this has been known as the Thomas–Kuhn–Reiche sum rule and, as we will see, it played a key role in creation of quantum mechanics.

The agreement of the oscillator model with experiment suggests that we should take seriously that the electron bound in the atom is moving under influence of the elastic forces

$$
\vec{F} = -m \sum_k f_k \omega_k^2 \vec{q}.
$$

Now, according to the classical theory, the electron in the atom is moving in the electrostatic field created by the other electrons and by the nucleus. If we substitute for the charge density from Eq. (2.3) into the solution for scalar potential,

Eq. (2.23), we get from Eqs. (2.4) and (2.12) that the electron actually moves under the influence of the force

$$\vec{F} = -e\nabla\varphi = -e\nabla\left(\sum_n \frac{e_n}{4\pi} \frac{1}{|\vec{q} - \vec{q}^{(n)}|}\right).$$

So we see the oscillator model does not make sense at all. In fact, the classical theory is not able to describe either the stability of the atom or its interaction with radiation. Nonetheless, the agreement of the oscillator model with experiment shows that there is a classical way to represent the atom–radiation interaction. The oscillator model provides a bridge from the classical theory to the correct theory that governs the motion of the electron in atom and its interaction with radiation—the quantum theory.

Chapter 3
Hamiltonian Formulation of Classical Mechanics

Starting from the Hamilton variational principle, we derive here the Hamilton canonical equations, the canonical transformations, and the action–angle variables. This is again the standard material that can be found in many places, see, e.g., [7]. The notions of the Hamilton canonical equations, the Hamilton's function, "Hamiltonian," and the canonical coordinates and momenta are crucial for transition from the classical to the quantum theory. This is true not only for understanding the historical development of quantum theory. These notions form the basis of the procedure of the canonical quantization that we use to this day. For understanding the historical development of the quantum theory, the notion of the action–angle variables is crucial too. We shall use it extensively in the following two chapters. However, its significance is only historical, and nowadays, the notion of the action–angle variables is almost forgotten.

3.1 Hamilton Variational Principle

As discovered by William R. Hamilton, the classical mechanics can be formulated variationally. The requirement is to ensure that the variation of action S is zero. The action S is the functional of the canonical coordinates and momenta of the particles,

$$\delta S = 0, \quad S = \int_{t_I}^{t_F} dt \, [p_i \dot{q}_i - H(q_i, p_i, t)] , \tag{3.1}$$

subject to the boundary conditions of vanishing variations of coordinates and momenta at initial and final times,

$$\delta p_i(t_F) = \delta p_i(t_I) = \delta q_i(t_F) = \delta q_i(t_I) = 0 . \tag{3.2}$$

© The Author(s), under exclusive license to Springer Nature Switzerland AG 2023
J. Zamastil, *Understanding the Path from Classical to Quantum Mechanics*,
SpringerBriefs in Physics, https://doi.org/10.1007/978-3-031-37373-2_3

For the cases considered here, the Hamiltonian H has the meaning of the total energy of a system. Equation (3.1) yields

$$\int_{t_I}^{t_F} dt \left[\delta p_i \dot{q}_i + p_i \delta \dot{q}_i - \frac{\partial H}{\partial p_i} \delta p_i - \frac{\partial H}{\partial q_i} \delta q_i \right] = 0.$$

By means of the identity

$$p_i \delta \dot{q}_i = \frac{d(p_i \delta q_i)}{dt} - \dot{p}_i \delta q_i,$$

we transform the variation $\delta \dot{q}_i$ to the variation of δq_i. Thus, the requirement for the vanishing of the variation of the action can be written

$$\int_{t_I}^{t_F} dt \left[\frac{d(p_i \delta q_i)}{dt} + \left(\dot{q}_i - \frac{\partial H}{\partial p_i} \right) \delta p_i - \left(\dot{p}_i + \frac{\partial H}{\partial q_i} \right) \delta q_i \right] = 0. \qquad (3.3)$$

The first term on the left-hand side of the last equation vanishes due to the condition (3.2). Since the variations of the coordinates and momenta are independent, Eq. (3.3) then yields the Hamilton canonical equations

$$\frac{\partial H}{\partial p_i} = \dot{q}_i, \qquad -\frac{\partial H}{\partial q_i} = \dot{p}_i. \qquad (3.4)$$

For instance, when in Eq. (2.42), we neglect the particle friction due to radiation— as mentioned above, this can be done when the frequency of the external EM field does not approach the oscillator frequency—the Hamiltonian describing the system becomes

$$H = \frac{\vec{p} \cdot \vec{p}}{2m} + \frac{m\omega^2}{2} \vec{q} \cdot \vec{q} - e\vec{E}_\perp(t) \cdot \vec{q}. \qquad (3.5)$$

The Hamilton canonical Eq. (3.4) then yields Newton equation of motion, Eq. (2.42), as one can easily verify.

3.2   Canonical Transformations

The canonical coordinates and momenta are not defined uniquely. By performing the canonical transformations

$$Q_i = Q_i(\vec{q}, \vec{p}, t), \ P_i = P_i(\vec{q}, \vec{p}, t), \ H(\vec{q}, \vec{p}, t) = \mathcal{H}(\vec{Q}, \vec{P}, t), \qquad (3.6)$$

the requirement of vanishing the action S, expressed now in new coordinates and momenta Q_i and P_i,

$$\delta S = 0, \quad S = \int_{t_I}^{t_F} dt \left[P_i \dot{Q}_i - \mathcal{H}(Q_i, P_i, t) \right], \tag{3.7}$$

subject to the boundary conditions

$$\delta P_i(t_F) = \delta P_i(t_I) = \delta Q_i(t_F) = \delta Q_i(t_I) = 0, \tag{3.8}$$

yields the Hamilton canonical equations in terms of new canonical coordinates and momenta

$$\frac{\partial \mathcal{H}}{\partial P_i} = \dot{Q}_i, \quad -\frac{\partial \mathcal{H}}{\partial Q_i} = \dot{P}_i. \tag{3.9}$$

Comparison of Eqs. (3.1) and (3.7) yields

$$p_i \dot{q}_i - H(q_i, p_i, t) = P_i \dot{Q}_i - \mathcal{H}(Q_i, P_i, t) + \frac{dF}{dt}, \tag{3.10}$$

where F is an arbitrary reasonable function. Now, consider this function to be a function of the old coordinates and the new momenta,

$$F = S(q_i(t), P_i(t), t) - Q_i P_i. \tag{3.11}$$

Equation (3.10) then yields

$$p_i \dot{q}_i - H(q_i, p_i, t) = P_i \dot{Q}_i - \mathcal{H}(Q_i, P_i, t) + \frac{\partial S}{\partial q_i} \dot{q}_i + \frac{\partial S}{\partial P_i} \dot{P}_i + \frac{\partial S}{\partial t}$$
$$- \dot{Q}_i P_i - Q_i \dot{P}_i. \tag{3.12}$$

Requiring now that the terms proportional to $\dot{q}_i$ and $\dot{P}_i$ vanish yields

$$p_i = \frac{\partial S}{\partial q_i}, \quad Q_i = \frac{\partial S}{\partial P_i}, \quad \mathcal{H} = H\left(q_i, p_i = \frac{\partial S}{\partial q_i}, t\right) + \frac{\partial S}{\partial t}. \tag{3.13}$$

We can insist that in terms of the new canonical coordinates and momenta, Q_i and P_i, the Hamiltonian vanishes, $\mathcal{H} = 0$; the last equation then yields the Hamilton–Jacobi equation

$$H\left(q_i, \frac{\partial S}{\partial q_i}, t\right) + \frac{\partial S}{\partial t} = 0. \tag{3.14}$$

From a mathematical point of view, the foregoing considerations transform the solution of $2n$ ordinary differential equations (3.4) to the solution of a single partial differential equation (3.14).

3.3   Action–Angle Variables

The transformation to action–angle variables J_i, w_i is a special case of the canonical transformation. Denoting $P_i = J_i$, $Q_i = w_i$, Eqs. (3.9) and (3.13) read

$$\frac{\partial \mathcal{H}}{\partial J_i} = \dot{w}_i \,, \quad -\frac{\partial \mathcal{H}}{\partial w_i} = \dot{J}_i \tag{3.15}$$

and

$$p_i = \frac{\partial S}{\partial q_i}\,, \quad w_i = \frac{\partial S}{\partial J_i}\,, \quad \mathcal{H} = H\left(q_i, p_i = \frac{\partial S}{\partial q_i}, t\right) + \frac{\partial S}{\partial t}\,, \tag{3.16}$$

respectively. This time we insist on

$$\dot{J}_i = 0 \Rightarrow \mathcal{H} = \mathcal{H}(J_i)\,, \quad \dot{w}_i = v_i \Rightarrow w_i = v_i t + \varphi_i\,, \tag{3.17}$$

where clearly v_i are time-independent and are called characteristic frequencies. This introduction of the action–angle variables appears to be a pure formality, and to a large extent it is, but, as already mentioned above, it is indispensable for understanding transition from classical to quantum mechanics.

For the particle executing a one-dimensional oscillatory motion with period T, the action variable J is equal to

$$J = \int_0^T p\dot{q}\,dt\,. \tag{3.18}$$

To show that this is the correct formula for the action variable, let us calculate the change of the angle variable per oscillation:

$$w(T) - w(0) = \int_0^T \frac{dw}{dq}\dot{q}\,dt = \int_0^T \frac{\partial^2 S}{\partial q \partial J}\dot{q}\,dt = \frac{\partial}{\partial J}\int_0^T \frac{\partial S}{\partial q}\dot{q}\,dt = \frac{\partial}{\partial J}J = 1\,. \tag{3.19}$$

In the second equality, we substituted from Eq. (3.16), and in the third equality from Eqs. (3.16) and (3.18). The last equation is precisely the required Eq. (3.17) since

$$w = vt + \varphi \Rightarrow w(T) - w(0) = vT = 1\,.$$

Chapter 4
Steps to the Correct Solution

In this chapter, we trace the steps that eventually led to the discovery of quantum mechanics. We start in Sect. 4.1 by summary of three significant empirically valid formulae:[1] the Planck radiation law, the Rydberg–Ritz combination principle, and the Balmer formula for hydrogen spectral lines. These formulae were firmly experimentally established, meaning that nobody questioned their validity. However, in contrast to the dispersion theory summarized in Chap. 2 it has become gradually clear that there is no way to derive these empirical results from the classical theory. It was Albert Einstein who, in his 1916 derivation of the Planck radiation law, made the first significant step to the creation of mathematically precise and consistent quantum theory. He introduced the absorption and emission coefficients describing the reaction of the atom to the external electromagnetic field. This is described in Sect. 4.2. The next step was taken in 1921 by experimental physicists Rudolf Ladenburg, who succeeded in relating the Einstein absorption coefficients to the oscillator strengths of the classical theory. This is explained in Sect. 4.3. In 1924, Hendrik Kramers slightly extended the Ladenburg formula, and in the same year, Max Born provided the derivation of the Ladenburg–Kramers formula.[2] This development is followed in Sect. 4.4. These developments prepared the way for the creation of the quantum mechanics. The Ladenburg–Kramers–Born dispersion theory was further extended in joint paper by Kramers and Heisenberg. We do not follow this further extension here, as it is not necessary for our purposes. Nonetheless, Heisenberg's acquaintance with the dispersion theory led him to his decisive breakthrough, as will be shown in the next chapter.

---

[1] These were certainly not the only significant empirically valid formulae known at the time, but they were the ones significant for the development we follow here.

[2] The English translation of all the original papers referred to here can be found in the collection [2].

© The Author(s), under exclusive license to Springer Nature Switzerland AG 2023
J. Zamastil, *Understanding the Path from Classical to Quantum Mechanics*,
SpringerBriefs in Physics, https://doi.org/10.1007/978-3-031-37373-2_4

4.1   Empirically Valid Formulae

4.1.1   Planck Radiation Law

In 1900, Max Planck published a law describing spectral distribution of the mean energy density ρ of an electromagnetic radiation at thermal equilibrium with a reservoir at temperature β^{-1} (measured in electron volts)

$$\rho(\omega) = \frac{\omega^3}{\pi^2} \frac{1}{\exp\{\beta\omega\} - 1}, \tag{4.1}$$

where the radiation circular frequency ω is related to the frequency ν via

$$\omega = 2\pi\nu. \tag{4.2}$$

The formula (4.1) was known to fit the experimental data within experimental errors. Planck arrived at it by interpolating between the Wien formula valid, purely on empirical grounds, for high frequencies ($\beta\omega >> 1$) and the Rayleigh–Jeans formula derived from classical theory and valid for low frequencies ($\beta\omega << 1$). As the Planck formula represents the very beginning of the quantum theory, and each of its four derivation, Planck's in 1900, Einstein's in 1905 and 1916, and Bose's in 1925, represented significant advancement of knowledge, both the genesis of the formula and its significance for additional developments can be found in many places. Personally, we recommend the book [8].

4.1.2   Rydberg–Ritz Combination Principle

In 1908, Walther Ritz generalized earlier work by Johannes Rydberg that frequencies of spectral lines of any atom can be obtained by adding or subtracting known frequencies of the lines of the same atom,

$$\omega_{nm} + \omega_{mk} = \omega_{nk}. \tag{4.3}$$

The Rydberg–Ritz combination principle is easily understandable from Planck–Einstein–Bohr radiation condition

$$\omega_{nm} = E_n - E_m, \tag{4.4}$$

i.e., that an atom absorbs or emits radiation only at frequencies corresponding to the differences between energies of its stationary states. The important notion of stationary states was introduced by Bohr in his 1913 atomic model: in a stationary state, the atom somehow, contrary to the classical electrodynamics, see Eq. (2.32), does not radiate.

4.1.3   Balmer Formula for Hydrogen

In 1885, then 60-year-old professor at grammar school for girls Johann Balmer published formula for frequencies of the hydrogen spectral lines

$$\nu_{nm} = Rc\left(\frac{1}{m^2} - \frac{1}{n^2}\right), \quad m, n = 1, 2, 3, \dots, \tag{4.5}$$

where the Rydberg constant R multiplied by speed of light in vacuum c equals (with the accuracy of the spectroscopic measurements achieved in 1885)

$$Rc = 3.2894 \times 10^{15} \text{Hz}.$$

Balmer guessed his formula on the basis of four spectral lines (known to him through the work of Anders Angström): the red one at $\lambda_{32} = 6562.10 \times 10^{-10}$ meters, the cyan one at $\lambda_{42} = 4860.74 \times 10^{-10}$ meters, and two violet ones at $\lambda_{52} = 4340.1 \times 10^{-10}$ meters and $\lambda_{62} = 4101.2 \times 10^{-10}$ meters. The wavelength and frequency of radiation are related by formula $\lambda_{nm} = \nu_{nm} c$.

4.2   Einstein's Derivation of the Planck Radiation Law

Neither directly nor indirectly did Einstein contributed (...) to the discovery of matrix mechanics by Heisenberg. A. Pais [8]

As we shall see, this is simply not true.[3] One is even tempted to say that Einstein should be called the grandfather of quantum mechanics in both its matrix and wave incarnations.

The Planck radiation law, Eq. (4.1), describes spectral distribution of the mean energy density ρ of an electromagnetic radiation at thermal equilibrium with a reservoir at temperature β^{-1}. The example of such a radiation is a radiation from an electric stove, from a light bulb, from stars, or the cosmic microwave radiation background. But what the reservoir consists of and how is the thermal equilibrium between the reservoir and the electromagnetic radiation established? By posing this question and answering it, Einstein made the first significant step toward precise mathematical formulation of quantum theory.

---

[3] This does not diminish the author's admiration for [8].

Einstein assumed that the reservoir consists of atoms and the thermal equilibrium is established via atom–radiation interactions. Let us consider the two atomic levels, and let us denote their energies and occupation numbers E_k, E_n, and N_k, N_n, respectively; let us denote by ω the radiation circular frequency corresponding to the differences between energies of the stationary states, cf. Eq. (4.4),

$$\omega = E_n - E_k \,. \tag{4.6}$$

The number of the atoms at the state n increases if the atom in the state k absorbs radiation and decreases if the atom in the state n undergoes stimulated or spontaneous emission to the state k, so

$$\frac{dN_n}{dt} = (N_k B_{kn} - N_n B_{nk})\rho(\omega) - N_n A_{nk} \,, \tag{4.7}$$

where B_{kn}, B_{nk}, and A_{nk} are coefficients of the absorption, stimulated and spontaneous emission, respectively. Let us first note that in the absence of radiation

$$\frac{dN_n}{dt} = -N_n A_{nk} \,, \tag{4.8}$$

so the coefficient A_{nk} expresses the probability per unit time that the atom in the state n spontaneously emits EM radiation and goes to the state k. Likewise, the coefficient B_{kn} expresses the probability per unit time per energy density that the atom in the state k absorbs radiation of energy density $\rho(\omega)$ and goes to the state n. Now, at thermal equilibrium, the number of atoms at the state n first does not change in time, and second, it is related to the number of atoms at the state k according to canonical distribution, cf. Eq. (4.6),

$$\frac{dN_n}{dt} = 0, \quad N_n = N_k \exp\{-\beta\omega\} \,. \tag{4.9}$$

Now, from Eqs. (4.7) and (4.9), the Planck radiation law (4.1) follows, and for the emission and absorption coefficients, we get that

$$A_{nk} = \frac{\omega^3}{\pi^2} B_{nk} \,, \quad B_{kn} = B_{nk} \,. \tag{4.10}$$

Clearly, the emission and absorption coefficients are independent of both temperature and radiation energy density, and they are just the properties of the atoms. By these very simple considerations, Einstein made in fact not one, but two revolutionary steps toward precise mathematical formulation of quantum theory: first, one has to consider two-indexed quantities related to the transitions between two stationary states, and second, one is better to be content just with the probabilities of the transitions, not their complete determination. Ironically, the second step Einstein never accepted as definite truth, though he alone introduced it. But, as far as we know today, it is definite truth.

4.3 Ladenburg Theory of Dispersion

Ladenburg's work is an example of those cases in which somebody developed something important but did not find a real appreciation of it for a long time. I cannot remember at all that Ladenburg's formula was ever mentioned in the many discussions on quantum theory in Göttingen during the first few semesters I was there. It somehow had not made its way, in general, into the quantum theoretical thinking. P. Jordan [3]

Rudolf Ladenburg had the brilliant idea to relate oscillator strengths, Eq. (2.69), to the Einstein coefficients of absorption and emission, Eq. (4.7). He did so by calculating the total power absorbed by the atoms first in classical way, according to the oscillator model, and second in quantum way, according to the Einstein theory. To determine the power absorbed by the oscillator, we multiply Eq. (2.42) by velocity; we obtain the law of conservation energy in the form, cf. Eq. (2.34),

$$-\frac{d\mathcal{E}}{dt} = -\mathcal{P}_{abs} + \mathcal{P}_{em}, \quad \mathcal{E} = \frac{m}{2}\left(\dot{\vec{q}}\cdot\dot{\vec{q}} + \omega^2\vec{q}\cdot\vec{q}\right), \tag{4.11}$$

$$\mathcal{P}_{em} = m\Gamma\dot{\vec{q}}\cdot\dot{\vec{q}}, \quad \mathcal{P}_{abs} = e\vec{E}_0\cdot\dot{\vec{q}}\cos(\omega_e t),$$

where $\mathcal{P}_{em}$ is the emitted power, cf. Eqs. (2.35) and (2.38), and $\mathcal{P}_{abs}$ is the absorbed power. By substituting for the dependence of the induced radius vector from Eq. (2.45), we obtain for the averaged power absorbed by the oscillator per one oscillation

$$\overline{\mathcal{P}}_{abs} = \frac{e^2}{m}\frac{\vec{E}_0\cdot\vec{E}_0}{2}\frac{\Gamma\omega_e^2}{(\omega^2 - \omega_e^2)^2 + (\Gamma\omega_e)^2}. \tag{4.12}$$

The radiation acting on the oscillator is now not monochromatic, but thermal. We have to replace the radiation energy density, $\vec{E}_0\cdot\vec{E}_0/2$, by the averaged energy density $\rho(\omega_e)$ and integrate over all circular frequencies of the incoming radiation

$$\overline{\mathcal{P}}_{abs} = \frac{e^2}{m}\int_0^\infty d\omega_e \frac{\rho(\omega_e)\Gamma\omega_e^2}{(\omega^2 - \omega_e^2)^2 + (\Gamma\omega_e)^2}$$

$$= \frac{e^2}{4m}\int_0^\infty d\omega_e \frac{\rho(\omega_e)\Gamma}{\left[(\omega - \omega_e)^2\left(\frac{\omega_e + \omega}{2\omega_e}\right)^2 + \left(\frac{\Gamma}{2}\right)^2\right]}. \tag{4.13}$$

Apparently, all the contribution to the integral comes from the neighborhood of $\omega_e = \omega$. Thus, around this point, we everywhere replace ω_e by ω and extend the integral over whole real axis

$$\overline{P}_{abs} = \frac{e^2}{4m} \int_{-\infty}^{\infty} d\omega_e \frac{\rho(\omega)\Gamma}{\left[(\omega - \omega_e)^2 + \left(\frac{\Gamma}{2}\right)^2\right]} = \frac{2\alpha\pi^2}{m}\rho(\omega). \tag{4.14}$$

This formula was derived already by Planck in his 1900 derivation of his radiation law. Now, the power absorbed by the N_a active oscillators calculated in classical way have to equal the same quantity calculated in quantum way so

$$N_a \frac{2\alpha\pi^2}{m}\rho(\omega) = \omega N_0 B_{01}\rho(\omega). \tag{4.15}$$

On the right-hand side, we multiplied the rate of absorption on right-hand side of Eq. (4.7) by the absorbed energy ω to get the absorbed power. Finally, Ladenburg interpreted the number of active oscillators as the number of atoms in normal state times the oscillator strength,

$$N_a = N_0 f_1 \quad \Rightarrow f_1 = \frac{m}{2\alpha\pi^2}\omega B_{01}. \tag{4.16}$$

Ladenburg now assumed that analogous relations hold for the other oscillator strengths as well. He thus obtained for the polarizability, cf. Eqs. (2.69) and (4.16),

$$a = N\frac{2}{\pi}\sum_k \frac{B_{0k}\omega_{k0}}{\omega_{k0}^2 - \omega_e^2}. \tag{4.17}$$

By comparing the absorption and dispersion measurements for the hydrogen and alkalis, Ladenburg checked that this formula is in agreement with the experiment, so his general identification of the oscillator strengths,

$$f_k = \frac{m}{2\alpha\pi^2}\omega_{k0}B_{0k}, \tag{4.18}$$

has to be correct.

4.4   Kramers Theory of Dispersion and Its Born's Derivation

It reminded me of the game which children play, in which you hide something in the room and then the child walks around to find it, and you say, "There you are quite cold, now you are getting warmer and so on." And as soon as a paper got warmer, so to say, one had the impression that it was more satisfying. One would say, "All right, that helps."... Born's paper was pretty good; it gave a proof of Kramer's dispersion formula. Everybody had the impression that one must try in this direction. Some people would perhaps criticize, "After all it is not much more than Kramer's paper. Kramers did all the right things that did mean anything. But still it definitely goes in the right direction." So there was a clear progress, step by step, from the Ladenburg paper, Kramer's paper, Born's paper, then the paper of Kramers and myself; every paper got a little bit further in the right direction. W. Heisenberg [3]

4.4.1 Kramers Extension of Ladenburg Formula

Hendrik Kramers suggested an extension of Ladenburg formula for the case when the atom is initially in the excited state

$$a = N \frac{2}{\pi} \sum_{\tau > 0} \left[\frac{B_{n,n+\tau} \omega_{n+\tau,n}}{\omega_{n+\tau,n}^2 - \omega_e^2} + \frac{B_{n,n-\tau} \omega_{n-\tau,n}}{\omega_{n-\tau,n}^2 - \omega_e^2} \right]. \tag{4.19}$$

The Kramers formula was later derived by Max Born. In the paper, he explicitly thanked Werner Heisenberg for help with carrying out the calculation. Kramers and Heisenberg then further developed in a joint paper the Kramers dispersion theory. As already mentioned above, Heisenberg involvement with the dispersion theory led him to his decisive breakthrough. Thus, it is worth to follow Born's derivation in detail. Born's derivation represents a great advance because he first showed that the Ladenburg and Kramers formulae are not at all dependent on the oscillator model and second he showed that the Einstein coefficients are proportional to square of the quantum replacements of the electron's Fourier coefficients.

In the following, we first derive the averaged induced electron radius vector in classical theory and then outline Born quantization procedure.

4.4.2 Canonical Perturbation Method

For definiteness, let us restrict ourselves to the situation of hydrogen and alkali atoms; since then, the interaction of long-wavelength EM radiation with atoms can be described as the interaction of EM radiation with one (valence) electron. Generalization of the following considerations to the more complex atoms is straightforward: if one has to consider n electrons, then there are $3n$ pairs of canonical coordinates and momenta instead of 3 pairs and one has to sum over $3n$-tuples of integers instead of triples of integers.

Let $\mathcal{H}_0$ be the atomic Hamiltonian.[4] The great advantage of the following derivation is that one does not have to assume anything about its precise form. The only assumption is that the atomic Hamiltonian is of such a form that supports the so-called multiply periodic or conditionally periodic electron motion: the Cartesian coordinates $\vec{q}^{(0)}$ of the electron moving in the atom can be expanded into the Fourier series

$$\vec{q}^{(0)}(\vec{w}^{(0)}, \vec{J}^{(0)}) = \sum_{\vec{\tau}} \vec{q}_{\vec{\tau}}(\vec{J}^{(0)}) \exp\{2\pi i \vec{\tau} \cdot \vec{w}^{(0)}\}, \qquad (4.20)$$

where $\vec{\tau}$'s are vectors of integers and $\vec{J}^{(0)}$, $\vec{w}^{(0)}$ are the action–angle variables, cf. Eq. (3.15),

$$\frac{\partial \mathcal{H}_0}{\partial J_i^{(0)}} = \dot{w}_i^{(0)}, \qquad -\frac{\partial \mathcal{H}_0}{\partial w_i^{(0)}} = \dot{J}_i^{(0)}, \qquad (4.21)$$

where, cf. Eq. (3.17),

$$\dot{J}_i^{(0)} = 0 \Rightarrow \mathcal{H}_0 = \mathcal{H}_0\left(J_i^{(0)}\right), \quad \dot{w}_i^{(0)} = v_i^{(0)} \Rightarrow w_i^{(0)} = v_i^{(0)} t + \varphi_i. \qquad (4.22)$$

The electron radius vector is a real quantity

$$\vec{q}^{(0)} = (\vec{q}^{(0)})^* \quad \Rightarrow \quad \vec{q}_{-\tau}^* = \vec{q}_\tau. \qquad (4.23)$$

The implication follows from inserting the Fourier expansion, Eq. (4.20), into the reality condition.

In the presence of an EM wave, the Hamiltonian changes to

$$\mathcal{H}(\vec{w}^{(0)}, \vec{J}^{(0)}) = \mathcal{H}_0(\vec{J}^{(0)}) + \epsilon \mathcal{H}_1(\vec{w}^{(0)}, \vec{J}^{(0)}, t), \qquad (4.24)$$

where ϵ is a perturbation parameter, eventually set to one. The Hamiltonian describing the interaction of the atom and the long-wavelength EM field reads, see Eq. (3.5),

$$\mathcal{H}_1 = -e\vec{E}_0 \cdot \vec{q}^{(0)}(t) \cos(2\pi v_e t) = \qquad (4.25)$$

$$= -\frac{e}{2}\vec{E}_0 \cdot \sum_{\vec{\tau}} \vec{q}_{\vec{\tau}}(\vec{J}^{(0)}) \exp\{2\pi i \vec{\tau} \cdot \vec{w}^{(0)}\} \left[\exp\{2\pi i w_e\} + \exp\{-2\pi i w_e\}\right],$$

where, as before, $\vec{E}_0$ describes the polarization and the amplitude of the EM wave, and v_e is its frequency. In the last equality, we substituted for $\vec{q}(t)$ from Eq. (4.20)

---

[4] In the following, quantities with zero superscript refer to the quantities evaluated for free atom.

and defined

$$w_e = v_e t .$$

(4.26)

In the presence of the EM wave, the canonical equations (4.21) still hold,

$$\frac{\partial \mathcal{H}}{\partial J_i^{(0)}} = \dot{w}_i^{(0)} , \qquad -\frac{\partial \mathcal{H}}{\partial w_i^{(0)}} = \dot{J}_i^{(0)} ,$$

(4.27)

but Eq. (4.22) does not. Nevertheless, we can make another canonical transformation to a new pair of canonical coordinates and momenta, cf. Eqs. (3.13) and (3.16),

$$J_i^{(0)} = \frac{\partial S}{\partial w_i^{(0)}} , \quad w_i = \frac{\partial S}{\partial J_i} , \quad \mathcal{H}' = \mathcal{H} \left(w_i^{(0)} , J_i^{(0)} = \frac{\partial S}{\partial w_i^{(0)}} , t \right) + \frac{\partial S}{\partial t} ,$$

(4.28)

such that the Hamilton canonical equations

$$\frac{\partial \mathcal{H}'}{\partial J_i} = \dot{w}_i , \qquad -\frac{\partial \mathcal{H}'}{\partial w_i} = \dot{J}_i$$

(4.29)

and their solutions, Eqs. (3.17), now hold

$$\dot{J}_i = 0 \Rightarrow \mathcal{H}' = \mathcal{H}'(J_i) , \quad \dot{w}_i = v_i \Rightarrow w_i = v_i t + \varphi_i .$$

(4.30)

We search for the solution of Eq. (4.28) in the form of a power series in the parameter ϵ

$$S(\vec{w}^{(0)}, \vec{J}, t) = \vec{w}^{(0)} \cdot \vec{J} + \epsilon S_1(\vec{w}^{(0)}, \vec{J}, t) + \dots$$

(4.31)

and

$$\mathcal{H}'(\vec{J}) = \mathcal{H}'_0(\vec{J}) + \epsilon \mathcal{H}'_1(\vec{J}) + \dots .$$

(4.32)

Substituting the expansions (4.24), (4.31), and (4.32) into Eq. (4.28), we get, with the accuracy to the first order in ϵ,

$$J_i^{(0)} = J_i + \epsilon \frac{\partial S_1}{\partial w_i^{(0)}} , \qquad w_i = w_i^{(0)} + \epsilon \frac{\partial S_1}{\partial J_i} ,$$

(4.33)

$$\mathcal{H}'_0(\vec{J}) + \epsilon \mathcal{H}'_1(\vec{J}) = \mathcal{H}_0 \left(\vec{J}^{(0)} = \vec{J} + \epsilon \frac{\partial S_1}{\partial \vec{w}^{(0)}} \right) + \epsilon \mathcal{H}_1(\vec{w}^{(0)}, \vec{J}^{(0)}, t) + \epsilon \frac{\partial S_1}{\partial t} .$$

Comparing now the terms of the same order of ϵ in the last equation, one obtains successively

$$\mathcal{H}_0'\left(\vec{J}\right) = \mathcal{H}_0\left(\vec{J}\right), \quad \mathcal{H}_1'\left(\vec{J}\right) = \frac{\partial \mathcal{H}_0}{\partial \vec{J}^{(0)}} \cdot \frac{\partial S_1}{\partial \vec{w}^{(0)}} + \mathcal{H}_1\left(\vec{w}^{(0)}, \vec{J}, t\right) + \frac{\partial S_1}{\partial t}.$$

(4.34)

Substituting now from Eqs. (4.21), (4.22), and (4.26), we can rewrite the last equation

$$\mathcal{H}_1'\left(\vec{J}\right) = \mathcal{H}_1\left(\vec{w}^{(0)}, \vec{J}, w_e\right) + \vec{v}^{(0)} \cdot \frac{\partial S_1}{\partial \vec{w}^{(0)}} + v_e \frac{\partial S_1}{\partial w_e}.$$

(4.35)

This is the equation for two unknown functions, $\mathcal{H}_1'$ and S_1. Defining averaging over the angle coordinates $\vec{w}^{(0)}$ and w_e

$$\langle Q \rangle = \int_0^1 \ldots \int_0^1 d\vec{w}^{(0)} dw_e \, Q,$$

(4.36)

we first take advantage of the fact that by assumption $\mathcal{H}_1'$ is independent of the angle coordinates, so $\left\langle \mathcal{H}_1'\left(\vec{J}\right)\right\rangle = \mathcal{H}_1'\left(\vec{J}\right)$. Thus,

$$\mathcal{H}_1'\left(\vec{J}\right) = \left\langle \mathcal{H}_1\left(\vec{w}^{(0)}, \vec{J}, w_e\right) + \vec{v}^{(0)} \cdot \frac{\partial S_1}{\partial \vec{w}^{(0)}} + v_e \frac{\partial S_1}{\partial w_e} \right\rangle = 0.$$

(4.37)

The average of the first term on the right-hand side of the last equation vanishes because, cf. (4.25), $\int_0^1 \ldots d\vec{w}^{(0)} \exp\{2\pi i \vec{\tau} \cdot \vec{w}^{(0)}\} = \delta_{\vec{\tau}, \vec{0}}$. The averages of the second and third terms on the right-hand side vanish because of the implicit assumption $S_1(\vec{w}^{(0)}, \vec{J}, w_e)$ is periodic in the angle variables $\vec{w}^{(0)}$ and w_e, for instance, $S_1(w_1^{(0)} = 0) = S_1(w_1^{(1)} = 1)$ and so on. Further, substituting for $\mathcal{H}_1$ from Eq. (4.25), and using Eq. (4.37), we find that Eq. (4.35) has the solution

$$S_1 = \frac{e}{4\pi i} \vec{E}_0 \cdot \sum_{\vec{\tau}} \vec{q}_{\vec{\tau}}(\vec{J}) \exp\{2\pi i \vec{w}^{(0)} \cdot \vec{\tau}\} \left[\frac{\exp\{2\pi i w_e\}}{\vec{v}^{(0)} \cdot \vec{\tau} + v_e} + \frac{\exp\{-2\pi i w_e\}}{\vec{v}^{(0)} \cdot \vec{\tau} - v_e} \right].$$

(4.38)

In the presence of the EM wave, the electron coordinate changes to

$$q_k = q_k(\vec{w}, \vec{J}) = q_k^{(0)}(\vec{w}^{(0)}, \vec{J}^{(0)}) + \epsilon q_k^{(1)} + \ldots,$$

(4.39)

where we expanded the coordinate in a Taylor series around its original atomic value, Eq. (4.20)

$$\epsilon q_k^{(1)} = (\vec{w} - \vec{w}^{(0)}) \cdot \frac{\partial q_k^{(0)}}{\partial \vec{w}^{(0)}} + (\vec{J} - \vec{J}^{(0)}) \cdot \frac{\partial q_k^{(0)}}{\partial \vec{J}^{(0)}}. \tag{4.40}$$

Substituting into Eq. (4.40) from Eq. (4.33), we get

$$q_k^{(1)} = \frac{\partial S_1}{\partial \vec{J}}\bigg|_{\vec{J} = \vec{J}^{(0)}} \cdot \frac{\partial q_k^{(0)}}{\partial \vec{w}^{(0)}} - \frac{\partial S_1}{\partial \vec{w}^{(0)}}\bigg|_{\vec{J} = \vec{J}^{(0)}} \cdot \frac{\partial q_k^{(0)}}{\partial \vec{J}^{(0)}}. \tag{4.41}$$

Defining the atomic average

$$\langle Q \rangle_0 = \int_0^1 \dots \int_0^1 d\vec{w}^{(0)} \, Q, \tag{4.42}$$

we get, by substituting Eqs. (4.20) and (4.38) into the last two equations and using

$$\int_0^1 \dots \int_0^1 d\vec{w}^{(0)} \exp\{2\pi i (\vec{\tau} - \vec{\tau}') \cdot \vec{w}^{(0)}\} = \delta_{\vec{\tau},\vec{\tau}'}, \tag{4.43}$$

for the averaged induced electron radius vector

$$\left\langle q_k^{(1)} \right\rangle_0 = \frac{e[E_0]_j}{2} \sum_{\vec{\tau}} \vec{\tau} \cdot \frac{\partial}{\partial \vec{J}^{(0)}} \left[(q_{\vec{\tau}})_j (q_{-\vec{\tau}})_k \right] \left[\frac{\exp\{2\pi i w_e\}}{\vec{\nu}^{(0)} \cdot \vec{\tau} + \nu_e} + \frac{\exp\{-2\pi i w_e\}}{\vec{\nu}^{(0)} \cdot \vec{\tau} - \nu_e} \right]. \tag{4.44}$$

Now, a product of the components of any two vectors $\vec{A}$ and $\vec{B}$ can be decomposed into symmetric and antisymmetric parts. The antisymmetric part can be related to the vector product of the two vectors, and the symmetric part can be further decomposed into a scalar and a traceless tensor,

$$A_j B_k = \frac{1}{2}[A_j B_k + A_k B_j] + \frac{1}{2}[A_j B_k - A_k B_j] \tag{4.45}$$

$$= \left[\frac{1}{3}\delta_{jk}\vec{A} \cdot \vec{B} \right] + \left[\frac{1}{2}\varepsilon_{jkl}\varepsilon_{lmn} A_m B_n \right] + \left[\frac{1}{2}(A_j B_k + A_k B_j) - \frac{1}{3}\delta_{jk}\vec{A} \cdot \vec{B} \right].$$

In the last equality, we used the identity (1.3) for the antisymmetric part. As atoms are spherically symmetric, only the scalar part survives the summation over $\vec{\tau}$ in Eq. (4.44). Therefore, we can make replacement

$$(q_{\vec{\tau}})_j (q_{-\vec{\tau}})_k \rightarrow \frac{1}{3}\delta_{jk}\vec{q}_{\vec{\tau}} \cdot \vec{q}_{-\vec{\tau}}. \tag{4.46}$$

Further, the expression in the square bracket in Eq. (4.44) can be simplified to

$$\frac{1}{2}\left[\frac{\exp\{2\pi i w_e\}}{\vec{v}^{(0)}\cdot\vec{\tau}+\nu_e}+\frac{\exp\{-2\pi i w_e\}}{\vec{v}^{(0)}\cdot\vec{\tau}-\nu_e}\right]=\frac{\vec{v}^{(0)}\cdot\vec{\tau}\cos\{2\pi w_e\}-i\sin\{2\pi w_e\}}{(\vec{v}^{(0)}\cdot\vec{\tau})^2-\nu_e^2}.$$

Now, contribution of the imaginary part vanishes because it is odd in $\vec{\tau}$, cf. Eqs. (4.44) and (4.46). The averaged induced electron radius vector can then be brought into the final form

$$\left\langle\vec{q}^{(1)}\right\rangle_0=\frac{2e}{3}\vec{E}_0\cos\{2\pi w_e\}\sum_{\vec{\tau}>0}\vec{\tau}\cdot\frac{\partial}{\partial\vec{J}^{(0)}}\left[\frac{\vec{v}^{(0)}\cdot\vec{\tau}|\vec{q}_{\vec{\tau}}|^2}{(\vec{v}^{(0)}\cdot\vec{\tau})^2-\nu_e^2}\right], \tag{4.47}$$

where, cf. Eq. (4.23),

$$|\vec{q}_{\vec{\tau}}|^2=\vec{q}_{\vec{\tau}}\cdot\vec{q}_{-\vec{\tau}}. \tag{4.48}$$

The restriction $\vec{\tau}>0$ means that we leave out those triples of $\vec{\tau}$ that can be obtained from those left in the sum in Eq. (4.47) by the interchange $\vec{\tau}\rightarrow-\vec{\tau}$. Comparison of Eq. (4.47) with Eqs. (2.45) and (2.46) yields the classical expression for the polarizability

$$a=\frac{2Ne^2}{3}\sum_{\vec{\tau}>0}\vec{\tau}\cdot\frac{\partial}{\partial\vec{J}^{(0)}}\left[\frac{\vec{v}^{(0)}\cdot\vec{\tau}|\vec{q}_{\vec{\tau}}|^2}{(\vec{v}^{(0)}\cdot\vec{\tau})^2-\nu_e^2}\right]. \tag{4.49}$$

4.4.3   Born Quantization Procedure

Born noted that if he writes the action variables in the form

$$J_k^{(0)}(\mu)=n_k+\mu\tau_k,\quad 0\le\mu\le1, \tag{4.50}$$

where n_k are quantum numbers of a stationary state, the correct transition from the classical to the quantum theory[5] is accomplished by sort of averaging the classical expression over the parameter μ: integrating the classical expression over μ from 0 to 1 and dividing the result by 2π. For the transitions from the classical frequencies

---

[5] The old Bohr–Sommerfeld theory set $J_k^{(0)}=n_k$, but as shown by Born and Heisenberg this led to the discrepancy with experiment for the excited states of the helium atom, see [3].

$\vec{v}^{(0)} \cdot \vec{\tau}$ to the quantum circular frequencies $\omega_{\vec{n}+\vec{\tau},\vec{n}}$, one obtains in this way

$$\omega_{\vec{n}+\vec{\tau},\vec{n}} = \int_0^1 d\mu \, \vec{v}^{(0)} \cdot \vec{\tau} = \int_0^1 d\mu \, \frac{\partial \mathcal{H}_0}{\partial \vec{J}^{(0)}} \cdot \frac{d\vec{J}^{(0)}(\mu)}{d\mu} = \tag{4.51}$$

$$= \int_0^1 d\mu \, \frac{d\mathcal{H}_0(\vec{J}^{(0)}(\mu))}{d\mu} = \mathcal{H}_0(\vec{n}+\vec{\tau}) - \mathcal{H}_0(\vec{n}) \,.$$

In the second equality, we used Eqs. (4.21), (4.22), and (4.50). In the last equality, we used Eq. (4.50). The last equation is precisely the radiation condition, Eq. (4.4). Likewise, Born noted that if in the classical expression for the polarizability, Eq. (4.49), he makes first the replacement

$$\frac{\vec{v}^{(0)} \cdot \vec{\tau} |\vec{q}_{\vec{\tau}}|^2}{(\vec{v}^{(0)} \cdot \vec{\tau})^2 - v_e^2} \rightarrow 2\pi \frac{\omega_{\vec{n},\vec{n}-\vec{\tau}} |\vec{q}_{\vec{n},\vec{n}-\vec{\tau}}|^2}{\omega_{\vec{n},\vec{n}-\vec{\tau}}^2 - \omega_e^2} \tag{4.52}$$

and second sort of averages over the parameter μ, then he gets for the polarizability

$$a = \frac{1}{2\pi} \int_0^1 d\mu \, \frac{2e^2 N}{3} \sum_{\vec{\tau}>0} \vec{\tau} \cdot \frac{\partial}{\partial \vec{J}^{(0)}} \left[2\pi \frac{\omega_{\vec{n},\vec{n}-\vec{\tau}} |\vec{q}_{\vec{n},\vec{n}-\vec{\tau}}|^2}{\omega_{\vec{n},\vec{n}-\vec{\tau}}^2 - \omega_e^2} \right] \rightarrow$$

$$\rightarrow \frac{2e^2 N}{3} \sum_{\vec{\tau}>0} \left[\frac{\omega_{\vec{n}+\vec{\tau},\vec{n}} |\vec{q}_{\vec{n}+\vec{\tau},\vec{n}}|^2}{\omega_{\vec{n}+\vec{\tau},\vec{n}}^2 - \omega_e^2} - \frac{\omega_{\vec{n},\vec{n}-\vec{\tau}} |\vec{q}_{\vec{n},\vec{n}-\vec{\tau}}|^2}{\omega_{\vec{n},\vec{n}-\vec{\tau}}^2 - \omega_e^2} \right], \tag{4.53}$$

where the replacement

$$\int_0^1 d\mu \, \vec{\tau} \cdot \frac{\partial \Phi}{\partial \vec{J}^{(0)}} = \int_0^1 d\mu \, \frac{\partial \vec{J}^{(0)}}{\partial \mu} \cdot \frac{\partial \Phi}{\partial \vec{J}^{(0)}} = \int_0^1 d\mu \, \frac{d\Phi}{d\mu} \rightarrow \Phi(\vec{n}+\vec{\tau}) - \Phi(\vec{n})$$

has been used. Comparing now the Born formula, Eq. (4.53), with the Kramers formula, Eq. (4.19), one gets for the Einstein coefficients

$$B_{\vec{n},\vec{n}\pm\vec{\tau}} = \frac{4\alpha\pi^2}{3} |\vec{q}_{\vec{n}\pm\vec{\tau},\vec{n}}|^2 \,, \tag{4.54}$$

or switching to a more convenient notation, in which the stationary state is labeled by its principal quantum number only,

$$B_{nk} = \frac{4\alpha\pi^2}{3} |\vec{q}_{nk}|^2 \,. \tag{4.55}$$

Given the earlier Ladenburg identification, Eq. (4.18), the Thomas–Kuhn–Reiche sum rule for the oscillator strengths, Eq. (2.71), now reads

$$\frac{2m}{3} \sum_k \omega_{kn} |\vec{q}_{nk}|^2 = 1 ,$$
(4.56)

where n is the principal quantum number of an atomic stationary state and the sum over k is the sum over all atomic stationary states.

Chapter 5
Heisenberg's Magical Steps

I then regard him - apart from the fact that he is personally also a very nice human being - to be very important, even gifted with genius, and I believe that he will one day advance science greatly. Pauli to Bohr, 11 February 1924 [3]

In this chapter, we finally come to Heisenberg's magical steps that constitute birth of quantum mechanics.[1]

Werner Heisenberg studied physics in Munich in the years 1920–1923, under the tutelage of Arnold Sommerfeld. In Sommerfeld's seminar, Heisenberg met his lifelong friend and colleague Wolfgang Pauli. After finishing his studies, Heisenberg went to Göttingen, where he was assistant to Max Born. In the winter 1924–1925, he spent about half of the year in Copenhagen as an assistant to Niels Bohr. Thus he was student of the best of the older experts on atomic physics. He started to work in 1922 on helium atom and Zeeman effects in atoms, the two problems he repeatedly returned to in the following years. He started to proceed in the framework of the old Bohr–Sommerfeld theory of the atomic structure, but application of this theory to the two problems made clear to him that this theory cannot be correct, as it cannot be brought into agreement with observations. Further, he learned that to achieve agreement with experiment, no mechanical atomic model can be taken too seriously. The line of attack on the atomic problems at the years 1923–1925 was on one hand to use Bohr correspondence principle—one should take seriously only those quantum formulae that in the limit of large quantum numbers go over to the classical formulae—on the other hand to use some kind of discretization of classical mechanics as advocated by Born and used successfully by him to derive Ladenburg–Kramers dispersion theory. What Heisenberg achieved in the early summer of 1925 was to first make synthesis of these approaches and second to sharpen these two somewhat vague ideas into the precise mathematical theory.

---

[1] English translation of the Heisenberg's original paper, as well as the following papers of Born, Jordan, Heisenberg, Dirac, and Pauli, can be found in the collection [2].

© The Author(s), under exclusive license to Springer Nature Switzerland AG 2023
J. Zamastil, *Understanding the Path from Classical to Quantum Mechanics*,
SpringerBriefs in Physics, https://doi.org/10.1007/978-3-031-37373-2_5

5.1   Fourier Series and Its Quantum Reinterpretation

*Then I saw that the analogy [of the quantum theoretical quantities] to the Fourier
components in classical physics was really very close because not only the absolute
value [of the transition amplitudes], but even the phase must somehow be well
defined. (...) One could see that the Fourier components were the reality, and
not the orbits. So one had to look for those connections between the Fourier
components, which were true in classical mechanics, and to see whether or not
similar connections were true also in quantum mechanics—i.e., if one took, instead
of the Fourier components, the transition amplitudes of the real lines of atoms.*
W. Heisenberg [3]

When the Born's formula for Einstein B coefficients (4.55) is combined with
Einstein equation (4.10) relating his coefficients of spontaneous and stimulated
emission, one gets

$$A_{nk} = \frac{4\alpha}{3}\omega_{nk}^3|\vec{q}_{nk}|^2 . \tag{5.1}$$

As previously established in the preceding chapter, cf. Eq. (4.8), the coefficients A_{nk}
determine the the probability per unit time that the atom spontaneously emits the EM
radiation of the frequency ω_{nk} and goes from one stationary excited state labeled by
its principal quantum number n to a stationary state of lower energy labeled by its
principal quantum number k.

If one substitutes the Fourier expansion of the electron radius vector, Eq. (4.20),
into the classical formula for the radiated power, Eq. (2.32), evaluates the atomic
average, Eq. (4.42), and takes advantage of the formula (4.43), one gets for the
averaged radiated power

$$\langle \mathcal{P} \rangle_0 = \frac{2\alpha}{3}\sum_\tau (\vec{\omega}\cdot\vec{\tau})^4|\vec{q}_\tau|^2 . \tag{5.2}$$

Now, if the transition rate, Eq. (5.1), is multiplied by the energy ω_{nk}, one gets the
radiated power. Further, if one sums over all lower stationary states labeled by k,
one obtains for the total radiated power from the stationary state labeled by n

$$\mathcal{P}_{qu} = \sum_{k<n}\omega_{nk}A_{nk} = \frac{4\alpha}{3}\sum_{k<n}\omega_{nk}^4|\vec{q}_{nk}|^2 \simeq \frac{2\alpha}{3}\sum_k \omega_{nk}^4|\vec{q}_{nk}|^2 , \tag{5.3}$$

where the last equality presumably holds for highly excited states n.[2] Comparison of the last two equations reinforces what was present already in the Born replacement (4.52): to get from classical to quantum theory, one should make the replacement

$$\vec{\omega} \cdot \vec{\tau} \rightarrow \omega_{n,n-\tau} \quad |\vec{q}_\tau|^2 \rightarrow |\vec{q}_{n,n-\tau}|^2 . \tag{5.4}$$

But Born made this replacement only at the end of his calculation. Heisenberg realized that such a replacement should be made at the very beginning of any calculation. Thus, the first, and the greatest of the Heisenberg's magical steps, the one that set him on the right track, was to replace the classical Fourier series, Eq. (4.20),

$$\vec{q}(t) = \sum_{\vec{\tau}} \vec{q}_{\vec{\tau}} \exp\{i\,(\vec{\tau} \cdot \vec{\omega}t + \vec{\tau} \cdot \vec{\varphi})\} , \tag{5.5}$$

by its quantum reinterpretation

$$\hat{\vec{q}}(t) = \sum_{\vec{n},\vec{\tau}} (\vec{q})_{\vec{n},\vec{n}-\vec{\tau}} \exp\{i\,(\omega_{\vec{n},\vec{n}-\vec{\tau}}t + \varphi_{\vec{n},\vec{n}-\vec{\tau}})\} , \tag{5.6}$$

or in customary and more convenient notation

$$\hat{\vec{q}}(t) = \sum_{n,m} (\vec{q})_{nm} \exp\{i\,(\omega_{nm}t + \varphi_{nm})\} . \tag{5.7}$$

As this representation of the coordinate is a real quantity, we have, cf. Eq. (4.23),

$$(\vec{q})_{nm}^* = (\vec{q})_{mn} . \tag{5.8}$$

5.2 Multiplication of Two QM Fourier Series

I had not realized that those things which I had used in my paper were matrices. I have never taken a course on matrix theory. Of course, I knew how to solve linear equations in a trivial way as one learns in school, but I have never learnt the general scheme of matrices and I did not know that matrices can represent groups. So when I learnt from Born what I have done was really an example of matrix multiplication, I was very interested, but it was new to me. Then, of course, I started to read up on it and learnt it. W. Heisenberg [9]

---

[2] In the second form of Eq. (5.3), we sum over all stationary states k of *lower* energy than the reference state n, while in the third form of Eq. (5.3) we sum over *all* stationary states.

The second of Heisenberg's magical steps was to ask what the multiplication of the quantum reinterpreted Fourier series looks like. He insisted that the multiplication of two quantum reinterpreted Fourier series has to be again a quantum reinterpreted Fourier series, so it has to have a form, cf. Eq. (5.7),

$$\hat{\vec{q}}(t) \cdot \hat{\vec{q}}(t) = \sum_{n,k} (\vec{q} \cdot \vec{q})_{nk} \exp\{i\,(\omega_{nk}t + \varphi_{nk})\}\,. \tag{5.9}$$

This lead Heisenberg to define the multiplication of the quantum reinterpreted Fourier series, Eq. (5.7), with itself as

$$\hat{\vec{q}}(t) \cdot \hat{\vec{q}}(t) = \sum_{n,m,k} (\vec{q})_{nm} \cdot (\vec{q})_{mk} \exp\{i\,((\omega_{nm} + \omega_{mk})t + (\varphi_{nm} + \varphi_{mk}))\}\,. \tag{5.10}$$

Comparison of the last two equations yields

$$(\vec{q} \cdot \vec{q})_{nk} = \sum_{m} (\vec{q})_{nm} \cdot (\vec{q})_{mk}\,, \quad \omega_{nk} = \omega_{nm} + \omega_{mk}\,, \quad \varphi_{nk} = \varphi_{nm} + \varphi_{mk}\,. \tag{5.11}$$

Now, the second equation, the equation for the circular frequencies, is nothing but Rydberg–Ritz combination principle, Eq. (4.3). Heisenberg thus knew that he was on the right track. Subsequently, Born, and independently of him Paul Dirac, recognized the matrix definition in the Heisenberg quantum reinterpreted Fourier series, Eq. (5.7), and in Eqs. (5.9)–(5.11) the definition of the matrix multiplication.

5.3   Equations of Motion

Turning now to the dynamics, Heisenberg made his third magical step when he supposed that the quantum dynamics had formally the same form as the classical one,

$$m\ddot{\hat{\vec{q}}} = -\nabla V(\vec{q})|_{\vec{q}=\hat{\vec{q}}}\,. \tag{5.12}$$

This line of attack, to change the kinematics of motion, but leave the dynamics intact, was exactly what Bohr foreshadowed by his correspondence principle. In the limit of large quantum numbers, the quantum formulae for measurable quantities have to go over to the corresponding classical formulae. As the Planck–Einstein–Bohr radiation condition (4.4) and the very notion of stationary states made clear, the classical kinematics has to be abandoned. To make sure that the correspondence principle is always satisfied, it is better to change the kinematics only.

For the simplest conceivable non-trivial dynamical problem, the one-dimensional harmonic oscillator, the explicit form of equation of motion, reads

$$m\ddot{\hat{q}} = -m\omega^2\hat{q} \quad \Rightarrow \quad \sum_{n,k}(\omega^2 - \omega_{nk}^2)q_{nk}e^{i(\omega_{nk}t+\varphi_{nk})} = 0, \tag{5.13}$$

where we inserted the one-dimensional version of (5.7)

$$\hat{q}(t) = \sum_{n,k}(q)_{nk}\exp\{i(\omega_{nk}t + \varphi_{nk})\}. \tag{5.14}$$

Not much can be gained from Eq. (5.13) on the first sight. To determine the form of the solution, Heisenberg resorted to correspondence arguments.[3] The solution of the corresponding classical problem

$$\ddot{q} + \omega^2 q = 0$$

has the form

$$q(t) = A\cos(\omega t + \varphi) = \frac{A}{2}\left(e^{i(\omega t+\varphi)} + e^{-i(\omega t+\varphi)}\right).$$

Comparing this classical solution with the quantization prescription, Eq. (5.4), Heisenberg guessed that unless $k = n \pm 1$, the transition amplitudes and frequencies are zero

$$(q)_{nk} = \delta_{k,n+1}q_{n,n+1} + \delta_{k,n-1}q_{n,n-1}, \quad \omega_{nk} = -\omega\delta_{k,n+1} + \omega\delta_{k,n-1},$$

$$\varphi_{nk} = -\varphi\delta_{k,n+1} + \varphi\delta_{k,n-1}. \tag{5.15}$$

The identification $\omega_{n,n\pm1} = \mp\omega$ also follows from the equation of motion, Eq. (5.13).

5.4   Quantum Condition

If one finds a difficulty in a calculation that is otherwise quite convincing, one should not push the difficulty away; one should rather try to make it the center of the whole thing. (...) In my paper, the fact that XY was not equal to YX was very disagreeable to me. I felt this was the only point of difficulty in the whole scheme; otherwise, I would be perfectly happy. But this difficulty had worried me and I was not able to

---

[3] Nowadays, we know how to solve harmonic oscillator very efficiently, without any recourse to classical theory, see any textbook on quantum mechanics.

solve it. Only for the example of the anharmonic oscillator that I have worked out
was I able to get rid of this difficulty. I have written down the Thomas–Kuhn sum
rule as a quantization rule, but I have not recognized that this was just the pq-qp
condition. W. Heisenberg [9]

Quantum equations of motion, Eqs. (5.12) or (5.13), do not determine the
transition amplitudes q_{nk} and transition frequencies ω_{nk} uniquely. They have to be
supplemented by the quantum condition, the analog of the old Bohr–Sommerfeld
quantization rule. To find it, Heisenberg substituted the one-dimensional version of
the classical Fourier series (4.20),

$$q(t) = \sum_{\tau} q_{\tau} \exp\{i\,(\tau \omega t + \tau \varphi)\}\,, \tag{5.16}$$

and $p = m\dot{q}$ into the action integral, Eq. (3.18), and obtained

$$J = \int_0^T p\dot{q}\,dt = m \int_0^T (\dot{q})^2 dt = m2\pi \sum_{\tau} \tau^2 \omega q_{\tau} q_{-\tau} = m4\pi \sum_{\tau>0} \tau^2 \omega q_{\tau} q_{-\tau}\,. \tag{5.17}$$

Differentiating now both sides of equation with respect to the action variable, one
has

$$\frac{1}{2m} = 2\pi \sum_{\tau>0} \tau \frac{\partial}{\partial J} [\tau \omega q_{\tau} q_{-\tau}]\,. \tag{5.18}$$

Heisenberg first replaced the expression in the brackets according to the replacement
of the dispersion theory, Eqs. (4.52) and (5.4),

$$\tau \omega q_{\tau} q_{-\tau} \rightarrow \omega_{n,n-\tau} q_{n,n-\tau} q_{n-\tau,n}\,, \tag{5.19}$$

and then he used the Born quantization prescription, Eqs. (4.51) and (4.53),

$$\frac{1}{2m} = \sum_{\tau>0} \left[\omega_{n+\tau,n} q_{n+\tau,n} q_{n,n+\tau} - \omega_{n,n-\tau} q_{n,n-\tau} q_{n-\tau,n}\right] \tag{5.20}$$

or in more convenient notation, cf. Eq. (5.8),

$$\frac{1}{2m} = \sum_{k} \omega_{kn} |q_{kn}|^2\,. \tag{5.21}$$

This is precisely the one-dimensional version of Thomas–Kuhn–Reiche summation
rule, Eq. (4.56). Thus, Heisenberg knew that his fourth magical step was on the right
track, too.

Born, and independently of him, Dirac, noted that the Heisenberg quantum condition, Eq. (5.21), can be put into a somewhat more convenient form. If, instead of substituting $p = m\dot{q}$, one writes the Fourier series for the momentum, cf. Eq. (5.16),

$$p(t) = \sum_{\tau} p_{\tau} \exp\{i(\tau\omega t + \tau\varphi)\}, \tag{5.22}$$

and substitutes it together with the coordinate expansion, Eq. (5.16), into the action integral, Eq. (3.18), one obtains

$$J = \int_0^T p\dot{q}\,dt = 2\pi i \sum_{\tau} \tau q_{\tau} p_{-\tau}. \tag{5.23}$$

Differentiating now the both sides with respect to the action variable, one has

$$-i = 2\pi \sum_{\tau} \tau \frac{\partial}{\partial J}[q_{\tau} p_{-\tau}] = 2\pi \sum_{\tau>0} \tau \frac{\partial}{\partial J}[q_{\tau} p_{-\tau} - q_{-\tau} p_{\tau}]. \tag{5.24}$$

Using now again the Born quantization prescriptions, one gets first

$$q_{\tau} p_{-\tau} - q_{-\tau} p_{\tau} \rightarrow q_{n,n-\tau} p_{n-\tau,n} - q_{n-\tau,n} p_{n,n-\tau} \tag{5.25}$$

and then

$$-i = \sum_{\tau>0} \left[q_{n+\tau,n} p_{n,n+\tau} - q_{n,n+\tau} p_{n+\tau,n} - q_{n,n-\tau} p_{n-\tau,n} + q_{n-\tau,n} p_{n,n-\tau} \right], \tag{5.26}$$

or in more convenient notation

$$-i = \sum_{k} [p_{nk} q_{kn} - q_{nk} p_{kn}]. \tag{5.27}$$

Writing now the quantum reinterpretation of the Fourier series (5.22),

$$\hat{p}(t) = \sum_{n,k} (p)_{nk} \exp\{i(\omega_{nk} t + \varphi_{nk})\}, \tag{5.28}$$

and using Heisenberg prescription for multiplication of two reinterpreted Fourier series, Eq. (5.10), one gets for the commutation of the two quantum reinterpreted

Fourier series, Eqs. (5.14) and (5.28),

$$\hat{p}(t)\hat{q}(t) - \hat{q}(t)\hat{p}(t) = \sum_{n,k} (pq - qp)_{nk} \exp\{i(\omega_{nk}t + \varphi_{nk})\},$$

$$(pq - qp)_{nk} = \sum_{m} [p_{nm}q_{mk} - q_{nm}p_{mk}] \, . \tag{5.29}$$

Thus, as noted by Born and Dirac, the Heisenberg quantum conditions, Eqs. (5.21) or (5.27), are statements about the non-commutativity of the multiplication of quantum reinterpreted Fourier series for the coordinates and momenta. The quantum condition (5.27) yields the diagonal elements

$$(pq - qp)_{nn} = -i \, . \tag{5.30}$$

Born's assistant Pascual Jordan, and independently of him Dirac, then guessed that the non-diagonal elements vanish

$$(pq - qp)_{nk} = -i\delta_{nk} \, . \tag{5.31}$$

Subsequently, Born, Heisenberg, and Jordan, and independently of them Dirac, then guessed the three-dimensional generalization of the last equation

$$\left(p_i q_j - q_j p_i\right)_{nk} = -i\delta_{nk}\delta_{ij} \, . \tag{5.32}$$

Thus, what Born with help of Jordan in Göttingen, and Dirac in Cambridge recognized in Heisenberg's magical steps is that transition from classical to quantum theory is exceedingly simple: it suffices to replace the canonical coordinates and momenta by the matrices, or, more generally, operators, and impose on them the canonical commutation relation, Eq. (5.32), and that is it.

5.5   Existence of a Ground State

Back to Heisenberg's struggle in the early summer of 1925. He noted that for the case of the harmonic oscillator, the equation of motion, Eq. (5.13), supplemented by quantum condition, Eq. (5.21), still does not suffice for a complete determination of the amplitudes and frequencies. Substitution of Eq. (5.15) into Eq. (5.21) yields the difference equation

$$\frac{1}{2m\omega} = |q_{n,n+1}|^2 - |q_{n-1,n}|^2 \, , \tag{5.33}$$

where we used, cf. Eq. (5.8), $|q_{n,n-1}|^2 = |q_{n-1,n}|^2$. This is clearly a difference equation for $q_{n+1,n}$ that has a general solution

$$|q_{n,n+1}|^2 = \frac{n+1+\delta}{2m\omega}. \tag{5.34}$$

After some struggle, Heisenberg realized that he could determine the remaining constant δ by imposing the requirement that there is a ground state energy pertaining to $n = 0$, so

$$q_{-1,0} = 0 \quad \Rightarrow \quad \delta = 0 \quad \Rightarrow \quad q_{n,n+1} = \sqrt{\frac{n+1}{2m\omega}}. \tag{5.35}$$

The quantum reinterpreted Fourier series, Eq. (5.14), thus has in the case of the harmonic oscillator the form

$$\hat{q}(t) = \frac{1}{\sqrt{2m\omega}} \sum_{n,k} \left[\delta_{k,n+1}\sqrt{n+1}\exp\{-i(\omega t + \varphi)\} + \delta_{k,n-1}\sqrt{n}\exp\{i(\omega t + \varphi)\} \right]. \tag{5.36}$$

5.6 Energy Conservation

Inserting now this form of solution into the expression for the energy, cf. Eq. (4.11), Heisenberg found

$$\mathcal{E} = \frac{m}{2}\left[\dot{\hat{q}}(t)^2 + \omega^2\hat{q}(t)^2\right] = \omega\left(n + \frac{1}{2}\right), \tag{5.37}$$

where he took the squares of the quantum reinterpreted Fourier series according to his prescription, Eqs. (5.10) and (5.11). In more detail, for the quantum reinterpreted Fourier series (5.36), one has

$$\dot{\hat{q}}(t) = i\sqrt{\frac{\omega}{2m}} \sum_{n,k}\left[-\delta_{k,n+1}\sqrt{n+1}\exp\{-i(\omega t + \varphi)\} \right.$$

$$\left. + \delta_{k,n-1}\sqrt{n}\exp\{i(\omega t + \varphi)\}\right],$$

$$\hat{q}(t)^2 = \frac{1}{2m\omega}\sum_{n,k}\left[\delta_{k,n+2}\sqrt{(n+1)(n+2)}\exp\{-2i(\omega t + \varphi)\} + \delta_{k,n}(2n+1)+ \right.$$

$$\left. + \delta_{k,n-2}\sqrt{n(n-1)}\exp\{2i(\omega t + \varphi)\}\right]$$

and

$$\dot{q}(t)^2 = -\frac{\omega}{2m} \sum_{n,k} \left[\delta_{k,n+2}\sqrt{(n+1)(n+2)} \exp\{-2\mathrm{i}\,(\omega t + \varphi)\} - \delta_{k,n}(2n+1)+ \right.$$

$$\left. + \delta_{k,n-2}\sqrt{n(n-1)} \exp\{2\mathrm{i}\,(\omega t + \varphi)\} \right].$$

Heisenberg then considered the anharmonic oscillator and determined the energy up to the second order of perturbation expansion in the anharmonicity constant, for details, see [4]. He was especially pleased that his calculational scheme yielded the energies automatically independent of time. Hence, they were conserved, as they should be.

5.7   Subsequent Developments

Now the damned Göttingen people use my beautiful wave mechanics to calculate their trashy matrix elements. E. Schrödinger [10]

Born, Jordan, and Dirac reformulated the Heisenberg theory into the Hamiltonian form. As a result, while the original Heisenberg paper may be difficult for modern readers to comprehend upon first reading, subsequent papers are much more familiar. In their third paper on quantum mechanics, Born, Heisenberg, and Jordan were able to deduce the general solution for angular momentum in quantum mechanics. Later, Wolfgang Pauli was able to use his erudition and technical skills to prove that Heisenberg's calculational scheme yields the Balmer formula for hydrogen spectral line, Eq. (4.5).

Roughly, half a year after Heisenberg submitted his paper, Erwin Schrödinger provided a precise mathematical formulation for Louis De Broglie's idea that electrons behave as waves. Schrödinger showed that his equation for De Broglie waves also yields the Balmer formula for hydrogen spectral lines. Subsequently, Schrödinger himself and Pauli demonstrated that Schrödinger wave mechanics is completely equivalent to Heisenberg quantum mechanics. Schrödinger wave mechanics provides a coordinate representation for canonical commutation relations, Eq. (5.32), and is mathematically superior for actual calculations in atomic physics compared to the original Heisenberg scheme. With the help of Schrödinger wave functions, Heisenberg was finally able to solve the helium problem. However, Schrödinger hoped to interpret electrons as real waves and completely eliminate concepts alien to classical physics, such as stationary states and quantum jumps. This hope never materialized, and he was not entirely pleased with these developments.

All of the subsequent developments, including the Hamiltonian formulation of quantum mechanics, the general solution for angular momentum in quantum mechanics, the Pauli solution for hydrogen using symmetries, the Schrödinger equation and its solution for hydrogen, representation or transformation theory, the relation between Heisenberg and Schrödinger representations, and the quantum theory of many-body systems, among others, are well-covered in numerous textbooks on quantum mechanics and need not be repeated here.

Chapter 6
Reflections on the Quantum Mechanics and the Path Leading to Its Discovery

The only thing that we learn from history is that we learn nothing from history.
G. Hegel

It suffices to compare general atmosphere and way of thinking in the scientific community active in the first quarter of the twentieth century and today to see how profound Hegel's maxim is. That a real understanding was achieved in such a matter that so thoroughly resists any human understanding is a miracle and is great distinction of the two generations of physicists active in the first quarter of the twentieth century. Given the near impossibility of the task and its successful completion, one would expect that each next generation would try to emulate these two generations in their general approach to science, the organization of the scientific society, the evaluation inside the society, and so on. Strange as it sounds, this is not quite so. Below are some reflections, some of them further elaborating foregoing statement, on the quantum mechanics and path leading to its discovery:

- No Web of Science, Scopus, or other databases were available at the time. Thus, there was no mindless counting of publications and citations. Theoretical physicists were not recognized for their ability to produce large numbers of publications, to influence and persuade others, or to collaborate with others. They were recognized solely for their individual ability to formulate objective mathematical laws that govern the behavior of Nature at the atomic scale and to determine their consequences. All physicists involved wrote papers to communicate their progress on their attempts to unravel Nature's secrets, and not to win job or grant applications, or to conform to the "publish or perish" philosophy. It was this very good communication, through published papers and personal letters, between a small group of distinguished physicists, notably Bohr and Kramers in Copenhagen, Born, Heisenberg, and Jordan in Göttingen, Pauli in Hamburg, Planck and Einstein in Berlin, Sommerfeld in Munich, and Ehrenfest in Leyden, that led to the rapid and successful development of the modern atomic theory. It was by no means an exclusive society; if an outsider

J. Zamastil, *Understanding the Path from Classical to Quantum Mechanics*, SpringerBriefs in Physics, https://doi.org/10.1007/978-3-031-37373-2_6

did really important work, its value was immediately recognized and absorbed. This applies to Dirac in Cambridge, Fermi in Rome, Bose in Calcutta, de Broglie in Paris, and Schrödinger in Zürich.

- Although they were some fellowships or scholarships available at the time, there were no grant agencies as we know them today. Therefore, no grant applications with proposals for future work. How could anyone in then existing state of affairs write definite plans for any future work? Could they claim to discover quantum mechanics in 3 years? Or could they propose to apply the known principles of the Bohr–Sommerfeld theory, which had already been proven incorrect, to understand atomic phenomena?

- There were scientific journals available at the time, but they were not necessarily distinguished by prestige or impact factors. While there were editors, there were no referees in the modern sense. Regardless, what could a referee have written about Heisenberg's 1925 paper? They might have either given up from the outset and admitted that the content and significance of the paper was beyond their grasp. Alternatively, they may have written a standard referee report of the kind that "the paper ignores or criticizes nearly all the work done in the field so far," "the ideas are not very clear and not very well developed," definitely "not very much was achieved," also "the physical interpretation of the formalism author proposes is completely obscure," overall "it seems that the author himself does not fully understand what is he trying to do." Conclusion: "The paper cannot be accepted in its present form." This is not an exaggeration, as demonstrated by the rejection of Zweig's original paper on quarks by referees, which forever remained in the form of a CERN preprint.

- There were no PR departments at the Universities at the time. No media cover. Heisenberg handed his paper to Born to judge if it is worth of anything and went away for travels and holiday. Born read it, gave the paper his blessing, and sent it to Zeitschrift für Physik for publication, where it was accepted. Could the publication of one of the most important papers in history of physics to be a more mundane event?

- Without Heisenberg and Schrödinger, the field of atomic and molecular physics would be mired in endless guesswork based on experimental data and mechanical models that were known to be fundamentally flawed. It would be a qualitative science in which little, if anything, held true, and where many different schools of similarly incorrect thought would be in constant conflict, each claiming that the others were completely wrong. This is a familiar pattern in all areas of human endeavor where precise and controllable experiments are lacking—including some areas of physics, such as quantum gravity. However, it is clear that precise and controllable experiments alone are not enough. The scientific community must have the search for truth as its foremost and indeed only goal.

As important as Heisenberg's magical steps were, equally important was the fact that Pauli, Born, Jordan, and Dirac immediately recognized that they constituted the first step toward the correct solution of the problem and developed Heisenberg's ideas further. Pauli, in particular, deserves credit for deriving Balmer's formula for the hydrogen spectrum from the newly born quantum mechanics.

- How little empirical input was actually needed! Remarkably, aside from black-body radiation, the physical phenomenon that eventually led to a decisive breakthrough was that of light dispersion, the phenomenon responsible for light refraction, which was already known to the ancients.

  However, small the empirical input was, it was absolutely crucial for arriving at the correct solution. In this respect, the importance of Born and Heisenberg's calculation of the Rydberg states of helium [3] and its comparison with reliable experimental data cannot be overestimated. The comparison unequivocally showed that the Bohr–Sommerfeld theory is wrong.

  Another critical aspect that contributed to the success of Heisenberg's work was requirement for the theory to be internally consistent and sufficiently rigid. This factor, rather than the famous focus on observable quantities, was likely Heisenberg's primary concern and helped him stay on the right track. Pauli shared this concern and, after successfully deriving the Balmer formula for hydrogen, immediately turned his attention to the problem of hydrogen in crossed electric and magnetic fields. It was known that slowly changing the external fields could cause the allowed Bohr orbit of hydrogen to deform into a forbidden orbit that collides with the nucleus [3, 9]. Pauli convinced himself that application of quantum mechanics to the problem of hydrogen in crossed fields does not lead, in contrast to the Bohr–Sommerfeld theory, to any difficulty.

- Although very few experimental data permit simple theoretical interpretation that can be directly useful for the formulation of natural laws, there was a plenty of experimental data available at the time, much more than theoreticians were able to explain either before or after the advent of quantum mechanics. Great theoretical ideas come when a theorist has a wealth of empirical data to think about, not when a bunch of theoreticians are impatiently awaiting the result of a single experiment. By the way, even today, except for the textbook examples of the hydrogen atom and Zeeman effect in alkalis, the quantitative interpretation of atomic spectra requires a lot of dirty numerical work and is not an easy task.

- Part of Heisenberg's genius was his ability to learn and take the best characteristics from others. Although his teachers and friends may have had their shortcomings—Born being too formalistic, Bohr too vague, and Pauli too critical—Heisenberg was able to profit from Born's mathematical erudition, Bohr's physical intuition, and Pauli's clarity and honesty. Nonetheless, despite the strong influence of his teachers and the best of his friends, he found himself completely alone with his thoughts during the decisive moments of early summer 1925.

- Although this study did not intend to present any new information about the historical progression of events, one remark is worth noting. It is commonly stated that Heisenberg's quantum mechanics originated from the Bohr–Sommerfeld theory of atomic structure. However, as explained previously, Heisenberg's mechanics grew out of Einstein's 1916 derivation of the Planck radiation law and the Ladenburg–Kramers–Born–Heisenberg dispersion theory. Heisenberg's involvement with the Bohr–Sommerfeld theory only made it clear to him that a classical description of electron motion had to be entirely discarded. It was

Einstein's introduction of the coefficients of absorption and emission, the two-indexed quantities, that put the other pioneers on the right path. There was no way to arrive at the correct theory from the old Bohr–Sommerfeld theory.

- Quantum mechanics, together with Einstein's special and general theories of relativity, has emphasized the importance and irreplaceability of mathematically precise formulation of the theory. Unlike Newton's geometrical presentation of his mechanical laws and Maxwell's mechanical model for his field equations, no geometrical or mechanical picture can capture the essence of atomic phenomena. At its fundamental level, Nature seems to prefer the analytic mode of thinking over the geometrical and the precise mathematical formulae over intuitive pictures. While Einstein's special and general theories of relativity showcase the power of human deductive reasoning, quantum mechanics demonstrates human ability to comprehend, through higher mathematics, something that would otherwise be incomprehensible.

- Just to clarify for the uninitiated, all the formulae given in the text are correct. Specifically, the Ladenburg–Kramers–Born–Heisenberg theory of dispersion, the Ladenburg relation between oscillator strengths and Einstein coefficients, and the Born relation between Einstein coefficients and transition amplitudes are all correct. All ideas and formulae in Heisenberg's pioneering paper are correct. Despite working in uncertainty, by checking their ideas against experiment, the pioneers got everything right.

- To ask whether quantum mechanics will be proven wrong someday, one must first consider whether classical mechanics was proven wrong. Classical mechanics has certain assumptions about the nature of the objects it describes, such as the assumption that one can speak about a massive point and assign to it a definite position and velocity at every time instant. These assumptions are not satisfied in the microscopic world, so classical mechanics is not applicable in that domain. However, there is nothing inherently wrong with classical mechanics even today. In fact, if one were to calculate the motion of planets using quantum mechanics, the only feasible approach would be to use a semiclassical approximation to ensure that quantum corrections are negligible and then use classical mechanics. Therefore, instead of replacing classical mechanics with quantum mechanics, it is better to view quantum mechanics as an extension of classical mechanics that is applicable in the domain where classical mechanics assumptions are not valid. In the realm of mathematical theories of physical phenomena, there seems to be a totalitarian principle: if a mathematical scheme works, then it works completely with all its consequences, no matter how strange they may seem at first glance. Furthermore, there is no other scheme that is not mathematically equivalent and that works just as well.

Currently, quantum field theories, including the famous Standard Model, assume that spacetime is continuous even at the smallest scales. However, it is possible that at the Planck scale, where quantum gravity effects should come into play, spacetime may no longer be continuous. It is conceivable that quantum theory would still be valid even under these conditions, but there may be some assumptions about the nature of objects described by quantum theory that no longer hold at the Planck scale. Therefore, let us assume that there is a better theory, called quantum gravity, that transcends quantum theory in the same way that quantum theory transcends classical theory. If one were to calculate the hydrogen spectrum with quantum gravity, one would make a "semi-quantum" approximation to ensure that corrections due to quantum gravity are entirely negligible and then happily proceed within the framework of quantum mechanics. This is precisely the way that atoms are described within the framework of quantum electrodynamics, except that quantum electrodynamics corrections to quantum mechanical results are not entirely negligible, but only small enough to be neglected in the zeroth approximation. Therefore, one could safely say that, regardless of the future, atomic spectra will be calculated using quantum mechanics for as long as humanity remains interested in such pursuits. The same applies to the classical description of planetary motion. From this perspective, the experimental confirmation of Bell inequalities appears somewhat trivial, even though the design and execution of such experiments are certainly not trivial.

- The doubts expressed by Einstein about the internal consistency of quantum mechanics, which some smart people still hold to today (see, e.g., [11]), are subject to the same considerations. The so-called measurement problem (see, e.g., [11]) cannot be a real problem. If quantum mechanics were internally inconsistent, it would not work so well. While the Bohr–Sommerfeld theory had some limited empirical success, its validity could not be extended over the whole atomic domain. Quantum mechanics was developed to describe what happens at scales on the order of 10^{-10} meters. However, investigations of much smaller scales, down to distances of 10^{-20} meters, did not affect the basic mathematical structure of quantum theory. Moreover, phenomena such as superfluidity, superconductivity, and experimental demonstrations of increasingly large Schrödinger cats clearly indicate that there is no reason to expect quantum theory to break down at macroscopic, everyday scales.

- As mentioned in the Preface, Schrödinger's approach to quantum mechanics is easy to understand. De Broglie proposed to associate a wave with each particle. Schrödinger then wrote the equation that the De Broglie wave had to satisfy, and that is it. However, Heisenberg's formulation of quantum mechanics, especially as developed by Dirac, delves more deeply into the heart of the matter, at least from our current perspective. Schrödinger's formulation is more in the spirit of classical physics, but it is also much more superficial. Once one goes beyond a single spinless particle moving in a prescribed potential and deals with spin, two or more identical particles, quantum fields, and so on, any close analogy with classical physics must be given up. But how could anyone ever have come up with the idea of replacing coordinates and momenta with operators satisfying $[\hat{p}, \hat{q}] = -i$? Well, this is precisely what we attempted to answer in this essay.

References

1. M. Kundera, *Testaments Betrayed: An Essay in Nine Parts* (Harper Perennial, New York, 1993)
2. B.L. van der Waerden, *Sources of Quantum Mechanics* (Dover, New York, 1967)
3. J. Mehra, H. Rechenberg, *The Historical Development of Quantum Theory*, vol. 2. The Discovery of Quantum Mechanics 1925 (Springer, Berlin, 1982)
4. I.J.R. Aitchinson, D.A. MacManus, T.M. Snyder, Am. J. Phys. **72**, 1370 (2004)
5. R.P. Feynman, R.B. Leighton, M. Sands, *Feynman Lectures on Physics* (Basic Books, New York, 2011)
6. J. Schwinger, L. de Raad Jr, K.A. Milton, W. Tsai, *Classical Electrodynamics* (Perseus Books, New York, 1998)
7. H. Goldstein, C. Poole, J. Safko, *Classical Mechanics*, 3rd edn. (Pearson, New York, 2001)
8. A. Pais, *Subtle is the Lord, The Science and the Life of Albert Einstein* (Oxford University Press, Oxford, 1982)
9. J. Mehra, H. Rechenberg, *The Historical Development of Quantum Theory*, vol. 3. The Formulation of Matrix Mechanics and Its Modifications 1925–1926 (Springer, Berlin, 1982)
10. J. Mehra, H. Rechenberg, *The Historical Development of Quantum Theory*, vol. 5. Erwin Schrödinger and the Rise of Wave Mechanics (Springer, Berlin, 1987)
11. R. Penrose, *The Road to Reality: A Complete Guide to the Laws of the Universe* (Vintage, New York, 2007)

J. Zamastil, *Understanding the Path from Classical to Quantum Mechanics*,
SpringerBriefs in Physics, https://doi.org/10.1007/978-3-031-37373-2

Printed in the United States
by Baker & Taylor Publisher Services